OPERATIONISM

Publication Number 256
AMERICAN LECTURE SERIES

A Monograph in
The BANNERSTONE DIVISION *of*
AMERICAN LECTURES IN PHILOSOPHY

Edited by
MARVIN FARBER, Ph.D.
Department of Philosophy
University of Buffalo
Buffalo, New York

OPERATIONISM

By

A. CORNELIUS BENJAMIN

John Hiram Lathrop
Professor of Philosophy
University of Missouri
Columbia, Missouri

CHARLES C THOMAS • PUBLISHER
Springfield • Illinois • U.S.A.

CHARLES C THOMAS · PUBLISHER

BANNERSTONE HOUSE

301-327 East Lawrence Avenue, Springfield, Illinois, U.S.A.

Published simultaneously in the British Commonwealth of Nations by

BLACKWELL SCIENTIFIC PUBLICATIONS, LTD., OXFORD, ENGLAND

Published simultaneously in Canada by

THE RYERSON PRESS, TORONTO

Library of Congress Catalog Card Number: 55-7450

Printed in the United States of America

CONTENTS

OPERATIONISM

Chapter I

THE GENERAL PROBLEM
OF OPERATIONISM

1. Introduction

Ever since physicists experienced the shock of the relativity theory, which compelled them to abandon some of the most favored of their traditional "absolutes," they have been turning increased attention to a re-examination and criticism of their methods. One clear-cut indication of this new interest was the publication in 1927 of *The Logic of Modern Physics* by the Harvard University physicist, P. W. Bridgman. In this book a new theory of method was defined and proposed. In spite of Bridgman's insistence that he was merely describing the procedures of the physicists and not developing a new theory of knowledge, the point of view which he adopted has come to be called "operationism" or "operationalism." It claimed to provide a criterion for eliminating vague and meaningless concepts by arguing that all concepts should be defined in terms of empirically performable operations. While there was some criticism of Bridgman's formulation of the method, his general proposal was accepted by most physicists as the procedure most likely to contribute to the advancement of physics and, in particular, most likely to avoid future difficulties of the kind created by the theory of relativity.

While this fact would have been sufficient to justify the importance of Bridgman's contribution to the theory of method, a more significant consequence was the general adoption of his point of view by other scientists, particularly by sociologists and psychologists. This spread of interest in the method unfortunately resulted in a great confusion both in the understanding of what the operational method really involves and in the proper evaluation of the arguments which are raised for and against its use. There is need, therefore, for a critical examination of operationism. This should summarize the various contributions to the topic, distinguish and clarify the alternative definitions proposed, examine arguments for and against the use of the operational method, and provide a generalized operational theory.

2. Vagueness and Ambiguity of the Term

There are several difficulties confronting one who attempts to discuss operationism in a comprehensive and painstaking manner. Noteworthy among these is the vagueness and ambiguity of the term. Bridgman himself defines the word in a variety of ways. For example, the following statements, arranged in order of publication, define or otherwise significantly characterize his operational point of view: (1) *"The concept is synonymous with the corresponding set of operations,"* 1927 (*14:5*). (2) "Meaning . . . is to be sought . . . in operations," 1934 (*17:-103*). (3) Operations are a "necessary" but not a "sufficient" condition for the determination of meanings, 1938 (*21:116*). (4) "The operational aspect is not by any means the only aspect of meanings," 1952 (*29:257*). This change in the conception of the relation between meaning and operation amounts almost to a contradiction if statement (1) is compared with statement (4).

Interpreting Bridgman more liberally, we can readily detect in his writings at least two meanings of the term "operation" — one specific or narrow, and the other general or broad. The specific use restricts the word to physical operations, and even at times to metrical operations. The general use allows mental, verbal and "pencil and paper" operations to be included under the term. These two meanings are quite different and unless they are kept distinct great confusion results. Unfortunately, not only Bridgman but many of the other writers on operationism frequently use the term without further qualification, and the reader is often left more or less completely in the dark as to what positions these writers are defending in case they are accepting operationism, or as to what positions they are criticizing in case they are rejecting it.

Similar difficulties arise when other writers on the topic are taken into consideration. Each author seems to feel under obligation to define the term "operation," or to modify and improve upon definitions already proposed. The result is a wide range and variety of definitions, many of them inconsistent with one another. Several studies of operationism *(36, 39, 47, 48)* take the form of attempts at "operational definitions of the term 'operation.'" One of these *(48)* is designed to show the unreliability of the accepted definitions of "operation." In this study four definitions were selected, and a group of graduate students at Duke University were asked to state whether each of twenty phenomena was or was not operational according to each of the four definitions. While the definitions varied widely in reliability the answers in all cases thoroughly justified the conclusion that greater reliability is badly needed. The net results of all these studies are hardly encouraging either from the point of view of precision in the way in which the term "operation" is defined, or from the point of view of consistency among the many definitions proposed. As a consequence, the general discussion of the operational point of view poses unusual problems.

3. Need for Analysis of Presuppositions

A second difficulty in discussing "operationism" centers about the problem of whether the point of view is to be taken as a concrete description of actual techniques of knowing, or as a general theory of knowledge. Bridgman's statement on this issue is unequivocal. He has vigorously denied that he is attempting to set up a philosophical system or to develop an elaborate and profound new theory of the nature of meaning (21:114). He has cautiously avoided the words "operationism" and "operationalism." He is not, he insists, attempting to present us with anything esoteric (21:114). He is not telling us how we *ought* to think, either in physics or in any other field. On the contrary he is simply describing certain methods which physicists have found successful in dealing with their subject-matter. These men have been led, particularly in recent years, to adopt a more critical attitude toward many of their concepts, and to initiate a method which is designed not only to meet certain difficulties which have arisen but also to prevent the appearance of even more complicated problems of a similar kind in the future. Thus Bridgman is saying in effect, "We find ourselves in an awkward situation in physics. This seems to be the result of certain methods which we have more or less unconsciously employed. Let us try another approach—the operational method—and see where it leads us." While he suggests that the method might well be extended to other areas of thinking, particularly to the social sciences, he avoids, at least in his later writings, the dogmatic claim that meanings *must* be operational, or that meanings involve *nothing but* operations. He restricts himself to asserting that as a physicist he has found it unnecessary to be concerned with anything more than the operational aspects of meaning (*21*:116).

In the face of such unprepossessing claims one hardly knows how to proceed. If Bridgman is not founding an "ism" but is merely describing certain methods which are employed by physicists today and which are proposed for more extended and consciously controlled use in the future, one can only ask whether the use is actually such as Bridgman describes, and whether the proposal for the future will succeed in avoiding the difficulties which it is designed to eliminate. In this case the "analysis" of operationism will consist in talking with physicists and watching them to see whether they are in fact employing the method, and in urging them to extend its use in order to see whether this brings about the anticipated results.

But there are two factors in the situation which suggest that a deeper analysis of the position is called for. In the first place, while Bridgman has not himself endeavored to discover the postulates which underlie his point of view, there clearly *are* certain presuppositions upon which it rests. The disclosing of such postulates is a difficult task, and one should not condemn Bridgman because he has no taste for this sort of enterprise. In a very recent statement he expresses surprise because so much of the concern of others in dealing with operationism has been with abstract methodological questions and with the erection of philosophical systems rather than with following out "the more concrete and obvious leads" (*30*). One can well understand Bridgman's lack of interest in such problems. But the fact remains that if the position is to be clearly understood someone must do this job; methods of knowing are justified only to the extent to which they result in knowledge, and a discussion of the criteria of knowledge leads inevitably into broad philosophical issues.

In the second place, while Bridgman has been reluctant to admit that he is founding anything that can be called a "system" or a "philosophy," his followers have not always been so modest. In fact, if we start with Bridgman's initially simple formulation of the position, and then consider the modifications in this point of view as these have been proposed by his disciples and his critics we can see that he has really, perhaps unwittingly, established both a creed and a school, with all of the usual accompaniments of such social organization—emotional attachments, vigorous proselytizing, and even internal dissension based on differences in the interpretation of the master. This makes it even more urgent to examine the position from the point of view of its role in a general theory of knowing.

4. Use of Source Material

A third difficulty in discussing "operationism" lies in deciding how much space one should devote to a mere exposition of Bridgman's ideas. It would be absurd, of course, to discuss the position without making frequent references to Bridgman. But since his writings are available no extended description of his ideas seems called for; he is presumably in a position to state his point of view more accurately than anyone else. He says, in fact, that he has seldom seen a printed discussion of the method which he would accept as an accurate presentation of his own ideas. It would, therefore, be superfluous and possibly even unfair to Bridgman to attempt an exposition of his position.

On the other hand, if an exposition of Bridgman's point

of view could be so devised as to contribute to the clarification of the various threads and stages of his thought this would certainly be helpful. It was suggested above, for example, that there are certain apparent inconsistencies in Bridgman's point of view. May it not be possible, however, that these represent not inconsistencies, but differences in emphasis because of special audiences for which he was writing, or changes indicating growth in his point of view, or even a rendering explicit in his later writings of something that was merely implicit in his earlier ones? While this growth might not be actually traceable in his writings, such an analysis of his position would serve to explain the fact that certain criticisms of his earlier position are not valid against his later outlook. It would also help to explain the fact that his followers do not always agree as to what his position involves, since they may have caught him at different stages in his development.

Furthermore, an analytic exposition of Bridgman's ideas might enable us to detect the various threads which make up the texture of his position. Two of these would certainly be empiricism and pragmatism. These are to be distinguished not only in Bridgman's point of view, but in that of other writers on the subject. Certain of these, notably Lundberg (*60:89*) and Stevens (*83*), appear to identify empiricism and operationism. This could hardly be said to be incorrect since there are so many varieties of each of these two positions that some justification for the interpretation could almost certainly be found. For the same reason the identification of operationism with logical positivism and the entire unity of science movement could not be said to be wrong; Roback (*74*) and some of the contributors to the "Symposium on Operationism" (*86*) appear to do this; and Rapoport (72:viii) goes even further, presumably identifying "operationalism" with "the philosophy of science." For the purposes of the following discussion none of these identifications need be made. The term "operationism" has enough ambiguities in its own right, without allowing it to absorb all of those of closely associated positions. This is not to deny that operationism is part of the empirical tradition, or that it has intimate connections with logical positivism, the unity of science and with the philosophy of science generally. Even in selecting empiricism and pragmatism as the main threads of operationism one may run into terminological difficulties. James, for example, considers empiricism and pragmatism to be substantially the same. And even if they are not identified one must grant that they have close affiliations with each other. An attempt should be made to keep the two positions separate. Though operationism draws heavily both from empiricism and from pragmatism, it probably owes more to the former than to the latter; Bridgman, for example, makes explicit reference to empiricism but seems to be more or less unconscious of the pragmatic element in his position.

Finally, an analytic exposition of Bridgman would provide an opportunity to indicate the main historical factors which may have contributed to the emergence of the operational point of view. Since Bridgman makes very few references to the history of thought this orientation would be in no way a statement of the actual sources of his thinking. The attempt could be made, rather, to show that from the broad historical traditions of empiricism and pragmatism operationism draws important elements. While it is certainly misleading to say, as Feigl does (44), that logical empiricism and operationism are "simply empiricism brought up-to-date," nevertheless operationism could be better understood if placed in the context of the empirical attitude. And while Bridgman seldom refers to pragmatism as such, his point of view has much in common with the position of Peirce, James and Dewey, and could be more effectively described as an example of this particular kind of theory of knowledge.

5. Outline for Discussion

The assumption will be made throughout the following pages that there *is* a general point of view which can be called "operationism." This is not to imply either that the adherents of the position agree as to what its main tenets are, or that all of them see clearly what specific postulates underlie it. Unquestionably it is, at the very minimum, a methodology. But as we shall consider the position it involves implicit answers to many questions of the general theory of knowledge. Probably there is also even a metaphysics, or world-view, implied by operationism, but considerations of this sort will not be emphasized in what follows.

I shall begin by discussing, largely without critical comment, Bridgman's point (or points) of view. In order to avoid the charge of misinterpreting him I shall use quoted statements liberally. Chapter II will be devoted to the empirical aspects of his operationism, and Chapter III to the pragmatic aspects. Whatever originality is to be found in this treatment will consist in the attempted separation of the various threads and stages of his thought, and in the brief references to the possible historical origins of the movement. In Chapter IV, I shall consider operationism as it has been taken over, either uncritically or with modifications, by writers particularly in the social and psychological sciences. In Chapter V, I shall present what seem to be the main inadequacies of the operational point of view; these will be designed to show not the impossibility of operationism but the difficulties which arise when it is interpreted too narrowly. This will prepare the way for the presentation, in Chapter VI, of a generalized operational theory.

Chapter II

BRIDGMAN'S EMPIRICISM

1. The General Nature of Empiricism

B ROADLY CONCEIVED, empiricism is the doctrine that experience is the sole source and the sole guarantee of knowledge. In the empirical tradition the definition of the word "experience" has, of course, been constantly under dispute. If the term is interpreted sufficiently broadly all knowledge *must* come from experience, for there can be no other source. The phenomenologists, in fact, have insisted that philosophy should begin with this broad conception of experience in which anything that can be intended or meant is given in some sense. But the historical empiricists have tried, in opposition to the rationalists, intuitionists, authoritarians and supernaturalistic theologians, to restrict experience to a greater or lesser degree. This has usually involved, first, a positive affirmation as to what *is* contained within experience, and, second, a rejection of certain supposed sources, such as the *a priori*, intuition, and mystic revelation.

2. Reliance on Experience

This positive element Bridgman has repeatedly emphasized. At the very outset he frankly admits the empirical framework of his thinking. The great lesson of empiricism, he argues, is that new kinds and orders of experience are always possible (*14:2*). What we have already learned may be a guide, but it must be only a guide,

(13)

as to what the future will disclose. No area of experience can legislate over other areas; experience as it reveals itself is the final court of appeal. Particularly significant in this connection are the changes which we have been compelled to recognize in recent years as we have pushed experience into new domains—areas of very high velocities, very small particles, very high pressures and very great distances. What we are discovering is that things do not behave on the "periphery" of experience as they do at its core. We must be prepared for all kinds of surprises.

The consequences of this for our conceptual framework are highly significant. One of the common traits of our linguistic behavior is the tendency to extend words beyond the areas which they were originally designed to describe (20:7; 18). If we are not careful we shall be led to suppose that all of the original connotations of these words apply to the novel situations. But a properly conceived empiricism should have taught us better. If new experiences are continually emerging have we any right to assume that our concepts have applicability when they are thus extended? We know that heat does not apply to individual molecules and color does not apply to individual electrons. May it not be also that space and time do not apply to very small particles, and that length and mass do not apply to objects traveling with extremely high velocities?

The specific problem of this new empiricism arose in Bridgman's thinking in the attempt to explain what has happened in recent physics, particularly in connection with the theory of relativity. This theory, argues Bridgman, has compelled us to modify our views as to what constitutes useful concepts in physics. It has shown that many concepts which had hitherto been defined in terms of "properties" can no longer be so defined. His favorite example is Newton's concept of time as "absolute, true and mathematical" (*14*:3-9). We can have no assurance whatsoever that anything of this kind exists in nature, since all our techniques for detecting the flow of time are relative to instruments and processes. There is no sense in speaking of an absolute time when the only time we can know is relative. Other concepts of essentially the same kind are empty space, the inside of an opaque body, the inside of an electron, the absolute motion of a star, and the identity of an electron. A physics which contains concepts of this kind becomes purely abstract and as far removed from reality as the abstract geometry of the mathematicians, built wholly on postulates.

The procedure for eliminating such concepts and replacing them by others which are more useful is a very simple one. It is based on the fundamental claim of empiricism that concepts get their meanings ultimately in terms of actual experiences. Consider, for example, the concept of length. We know what we mean by the concept of length only if we have some method by which we can tell the length of any given object. At the level of the usual objects of experience this method is clear-cut and definite; it involves performing the physical operations of measurement. "The concept of length involves as much as and nothing more than the set of operations by which length is determined . . . *the concept is synonymous with the corresponding set of operations*" (*14*:5).

From the point of view of clarity and precision this statement of the operational theory could hardly be improved upon. It seems to say exactly what Bridgman intended it to say. Concepts are meaningless unless they can be tied to experience in some way. Without empirical contacts they run the risk of becoming lost in mere verbalisms; words become substituted for words, and the process either goes on endlessly or terminates in the appeal to non-empirical sources. The only alternative is to recognize that the ultimate route of clarification is through direct experience. A term is defined, he argues, when the conditions are stated under which one may use the term and when one may infer from the use of the term by his neighbor that the same conditions prevail (*86:246*). Ultimately this always involves activity of some sort, even if it is nothing more than placing oneself in a position such that his sense organs may be acted on by a certain object. In physics it commonly involves physical operations, especially creating and setting up various kinds of recording and measuring devices, and then reading the results.

From this it seems clear that the operational approach should enable us to handle the situation created by the advent of the theory of relativity. In terms of an operational theory absolute space, time and motion become meaningless. If concepts are defined operationally, and specifically in terms of actually performable operations, a concept can have no meaning when extended to areas in which the operations cannot be performed. There is always the possibility, of course, that new operations may be found in the new areas. But if the operations are different, the concepts must be different. Even in borderline cases, where alternative operations may be performed and may be demonstrated to produce the same results, it is often advisable to recognize the difference in the operations by inventing two concepts, corresponding to the respective operations (*14:23; 21:121*). It is never safe to assume that the operations will continue to produce the same results when extended to the new area (*86:247*). If we always replace the old concepts by new ones when the operations change, we can be sure that we shall never have to retract anything we have already said; for the old concepts will continue to apply to the former area, and the new concepts—being definable by means of performable operations in the new region—must continue to apply to this area (*19:10*).

It is important, furthermore, that the operations be actually performable. This gives the concept the "validity of actual experience" (*86:246*), and prevents inconsistency or contradiction from entering into our concepts.

But how can we know what operations are actually performable? Definition will not enable us to decide, since in order to define an operation we must use another operation, and this must in turn be defined operationally. Bridgman avoids this infinite regress by pointing out that the verbal process eventually terminates in a situation where we can point at an operation and imitate it (*86:248*). Such an operation is closer than other forms to "primary experience" (*21:118*). No operation can, of course, be uniquely defined any more than can any other physical entity, and we can never be absolutely sure that we are correctly performing an assumed operation. But certain operations seem to us so simple that when we are told to perform them we can do it unequivocally (*21:118*).

3. Denial of the *A Priori*

It was stated above that empiricism usually involves as a positive doctrine a statement of what is contained within experience, and as a negative doctrine a rejection of other supposed sources of knowledge. We have seen something of Bridgman's positive doctrine. Negatively, also, however, he identifies himself with the empirical tradition.

In the first elaboration of his position (*14*:3) he makes this quite evident. "The attitude of the physicist must . . . be one of pure empiricism. He recognizes no *a priori* principles which determine or limit the possibilities of new experience. Experience is determined only by experience. This practically means that we must give up the demand that all nature be embraced in any formula, either simple or complicated." Since new experiences are always possible, and since our previous exploratory acts have disclosed the fact that as we extend experience we are continually being confronted with new objects and modes of behavior, we must rigidly refrain from legislating in advance as to the character of nature. All that we can do is to wait until these new areas have disclosed themselves to us; then we must take them as we find them.

In this strong rejection of the *a priori* Bridgman places himself firmly within the empirical tradition. Locke, too, rejected anything which presumed to be determinative or regulative of experience; Hume argued that no source other than experience would ever enable us to infer an effect from a cause; and Mill insisted that the uniformity of nature and the law of universal causation were not *a priori* beliefs but simply inductions from experience. Without doubt Bridgman would, in common with other empiricists, reject Kant's rationalistic claim that the propositions of mathematics are synthetic *a priori* judgments, though Bridgman's view as to the status of mathematics is not clear. On the one hand, he says that it is "just as truly an empirical science as physics or chemistry" (*19*:52); on the other, he insists that it is a "human invention" (*14*:60). Certainly these two statements are not equivalent to one another, and they might even be considered incompatible. But even if they cannot be reconciled they have in common the opposition to anything like an *a priori*; if mathematics is either a natural science or a human invention it will most certainly have nothing to do with innate ideas, or with principles which are self-evident or in any sense presupposed by experience. Thus one can say that Bridgman is an empiricist in his rejection of the *a priori*, though one might have difficulty determining from his writings just how he takes care of the problems which the conception of the *a priori* was designed to solve.

4. Rejection of Mysticism

As for mysticism, Bridgman is willing to grant that mystical definitions may be operational. But they are quite different from the operational definitions of the physicist. "Observation will show that the operations of the mystic include arguing for some system of philosophy" (20:77). "The state of mind that uses mystical definitions and is so convinced of their correctness that it is sure that no experimental check is necessary . . . has been shown by the experience of the last 300 years to be so hopeless for dealing with the world that we shall simply ignore it" (20:28). "All of our concepts of a mystical or supernatural character take their rise from the inertia of verbal patterns and forms, and . . . observably the only meaning which can be associated with these concepts is that they permit the operation of being used with certain verbal forms, that is, these are verbal concepts" (20:89).

Mysticism, therefore, is identified in Bridgman's mind with all philosophical system-building which uses rational as opposed to empirical methods. To characterize it as operational is not to justify it, for the operations are verbal rather than physical, and it is therefore "hopeless" for dealing with the world. One cannot assume, of course, that in rejecting mysticism as a method for physics he is also rejecting it as a method for religion or for some other phase of life; this question has not, so far as I know, been considered by him, and no answer can be assumed. But one can say that in other respects he gives every indication of identifying himself with the naturalistic strain of empiricism.

5. Restriction to Particulars

Thus Bridgman has explicitly stated that experience is to be understood in such a way as to exclude the *a priori* and the mystical. He has also, by implication, introduced a further restriction. Experience, for Bridgman, at least to the extent to which it is made up of operations, contains no universals but only *particulars.* "Operations are performed by human beings in time and are subject to the essential limitations of the time of our experience—the full meaning of any term involves the addition of a date—future operations mean nothing except as they are described in terms of operations performed now" (19:41). This means that an operation is an act performed by a given individual at a given time and place. As such it is not, strictly speaking, repeatable; another operation, however similar to the first, and even if performed by the same individual, would be a different operation. If I measure the length of a table twice, I cannot give the same meaning to the word "length" in the two cases since, even if I come out with the same result on each occasion, the two operations must differ at least as spatio-temporal events. Hence in this aspect of his position Bridgman may be taken to be denying all generality in experience. While it is true, as we shall see later, that he allows for *repeatable* operations and for operations which are *equivalent* to one another; nevertheless even these operations are particular, and he thus creates for himself the problem of accounting for abstract and general ideas, and of explaining why our vocabulary is not made up wholly of proper names.

This emphasis, again, represents an important phase of the empirical tradition. The empiricist has repeatedly argued that common sense is essentially right in claiming that knowledge is primarily of particular objects—the concrete things of our everyday lives, such as tables and chairs, rocks and mountains, plants and animals. These are characterized by spatio-temporal location, and each object is unique in the sense that two objects cannot occupy the same space at the same time. Space and time are principles of individuation, and all that we can know is specifically determined in this way. The negative equivalent of this is the denial of any such thing as a universal or a general object; there are only particular men, and there is no universal man.

This emphasis on particulars has been characteristic of empiricism, and has associated the position with all forms of nominalism and conceptualism. Some of the early Greek philosophers had theories of perception, in which the attempt was made to show how we obtain knowledge of particulars. While Plato departed from the empirical tradition by arguing that the forms were more real than the particulars, Aristotle is sometimes considered to have corrected his teacher in this regard. The emphasis on particulars again appears in the late Medieval period in William of Occam, and is strongly represented in the empiricisms of Hobbes, Locke and Hume. For all these writers universals have only the reality of ideas or of names, and there is needed a theory of abstractive operations by which concepts may be derived from individual objects. John Stuart Mill carried on the tradition and developed a theory of "comparison and abstraction" by which general ideas are extracted from particulars. Some of the operationists, as we shall see later, have tried to counteract this extreme emphasis on particulars by introducing "generalizing" operations; but most of the group either seem to be completely oblivious to the problem of the nature of general ideas, or to have decided that since only particulars exist the entire problem is an illegitimate one.

6. Conflict Between Objectivism and Relativism

There is still another aspect of Bridgman's position which suggests that it is part of the long tradition of empiricism. This factor is particularly significant since it represents an unresolved conflict in the empirical point of view—yet a conflict which has been present throughout the history of the movement and has exhibited itself as an alternation of views within the movement itself. This is the conflict between objectivism and relativism. On the one hand, the empiricists have tended to think of themselves as realists; many have even accepted materialism, and most empiricists have felt that in knowing we are presented with a world which exists independently of us and which we are able to know in its essential characteristics. If this is true, knowing is primarily passive rather than active, it "does nothing" to the object but merely receives impressions from it. One thinks immediately of Bacon's "listening to nature," and of Locke's "passive understanding" which merely records impressions upon a "white paper, void of all characters." Knowing is not a matter of creating, or even of modifying, but merely of receiving. But, on the other hand, many empiricists have recognized that the situation is more complex than this, and that the perceiver may, in the very act of perception, not only contribute something to the known object but even actually create it. The result is that our knowledge may be considered to be a function not merely of the object known but also of our knowing operations. Some such view as this is found as early as Empedocles. According to him knowledge of an object is produced by the meeting of particles from the object and particles from the eye; the image which results from this meeting is in part created by the observer since his sense organs are active rather than passive in the process and thus contribute to the character of the knowledge. Hence knowledge is relative, not objective, and we may never be able to know objects as they really are.

In Bridgman these two conflicting strains are equally evident. On the one hand, he seems to be an objectivist. "The only possible attitude toward the facts of experience as it unrolls is one of acceptance" (19:15). Experience is determined only by experience; as new areas are disclosed we can only accept what we find. Our present quandary in physics is the result of trying to force experience into molds determined by our own thinking. That what we are obliged to accept is the *physically* real, is perhaps not essential to Bridgman's position, although the main thread of empiricism strongly favors this interpretation; in view of Bridgman's later broadening of his point of view this restriction should not be insisted upon. But even if experience is extended to include the social as well, the final court of appeal is something which we do not ourselves make; it is something which is given to us and to which our conceptual tools must be adapted.

However, there is also in Bridgman a dominating relativism, which at many points becomes an out-and-out
subjectivism. One of the main consequences of the operational point of view, he argues, is that all knowledge is
relative (*14:25*), and one of the most revolutionary in-
sights which has come out of our recent experience in
physics is that we cannot transcend the human reference
point (*28:5*). This is true in a general sense; since knowledge rests upon operations and all operations are relative,
all knowledge must be equally so. But it is true also in a
special sense; since size, motion, rest and other concepts
are defined by operations which are specifically relative,
there can be no such thing as absolute size, or motion, or
rest. In a much more extreme sense of relativity the observer can never get outside himself at all; direct experience embraces only the things in his own consciousness—
sense impressions and various sorts of cerebrations—and
nothing else (*19:13*). This, Bridgman acknowledges, may
be solipsism, but if it is "we have got to adjust our thinking
so that it will not seem repugnant" (*19:15*). (It should be
noted in this connection that Alpert says flatly that operationism must be rejected if it requires a solipsistic epistemology (*4:857*)). That Bridgman cannot get away from
this subjectivism is indicated by the repeated references
which he makes to it, particularly in his later writings
(*20:142-59*; *26:ch. 3*; *86:281-4*). Yet he nowhere succeeds
in reconciling this with his objectivism.

It might appear that in the notion of operation Bridgman has, in fact, achieved this reconciliation of relativism
and objectivism. Consider the sense in which the meaning of any concept is relative to operations. This seems to
destroy objectivity, for we can know objects only by performing operations on them; and since operations are our

acts, and our acts are relative to us as actors, objects can be known only as they are known relative to us. But a kind of objectivity remains, for while it is true that operations are individual and performed at a specific time and place, they are repeatable and their relativity is thus not too significant. Furthermore, operations which are significantly different in kind may often be substituted for one another, hence the specific character of the operation is not always determinative of the resulting concept. Finally, the operations, at least in Bridgman's earlier statement of his position, are physical and thus have a distinctive kind of objectivity; for example, a concept which is operationally defined in terms of a physically impossible operation must be a meaningless one (*86:246-7; 19:9*)—a fact which prevents us from being carried away by operations and from introducing into our conceptual scheme notions which have no direct empirical relevance.

But that this does not solve Bridgman's problem can be seen by comparing his solution with that of Locke. If one disregards the difference between physical and mental operations the close resemblance between the views of these two individuals becomes at once apparent. Consider, for example, the position of Locke. While he argued for the essential passivity of man so far as simple ideas are concerned, he was convinced that knowledge contains also complex ideas. "As the mind is wholly passive in the reception of all its simple ideas, so it exerts several acts of its own, whereby out of its simple ideas as the materials and foundations of the rest, the others are framed. The acts of the mind, wherein it exerts its power over its simple ideas, are chiefly these three: (1) Combining several simple ideas into one compound one, and thus all *complex ideas* are made. (2) The second is bringing two ideas, whether simple or complex, together, and setting them by one another, so as to take a view of them at once, without uniting them into one; by which it gets all its *ideas of relations*. (3) The third is separating them from all other ideas that accompany them in this real existence; this is called abstraction: and thus all its *general ideas* are made" (55:213-4).

The close similarities between Bridgman and Locke are obvious. Both use the term "operation" to designate the acts which are required to give content to certain ideas. Both think of the operations as in some significant way involved in the determination of the meaning of the resultant concept. Both distinguish kinds of operations and endeavor to show how a derived concept is functionally related to the operation performed. Both would argue against the passivity of the knower, at least in this kind of knowing, and would insist that knowing always involves doing something, *i.e.*, performing certain acts.

But aside from these similarities there is an important difference. For Locke knowledge is not only of complex ideas, determined operationally, but also of simple ideas, not determined operationally. Thus Locke allows for the fact that we can know some things without the intermediary of operations, hence not all knowledge is relative. Bridgman, in insisting that all physical concepts be defined operationally, leaves no room for physical knowledge which can be acquired in any other way. Because of this we have no place to "attach" our operations, and we have no means of determining what it is that we are trying to know by means of operations. There exists for Locke a world of things with their primary and secondary qualities; these objects exist independently of being known, and can, moreover, be known either passively, by means of simple ideas and without the intermediary of operations, or actively, by means of complex ideas which require operations of the mind. For Bridgman there is no assurance of such a world, since, if it existed, we could know it only through operations and these would appear to determine its character exhaustively. We should carefully avoid the assumption, he insists, that because there are limits to the operationally meaningful there is a certain realm—the operationally meaningless—beyond this. For such a supposed world would readily "become the playground of the imagination of every mystic and dreamer" and "let loose a veritable intellectual spree of licentious and debauched thinking" (15:451).

7. Positivism

Finally, there is in operationism a strong positivistic element. This is more evident in some of Bridgman's followers, as we shall see in Chapter IV, than it is in Bridgman himself. It is expressed in an anti-metaphysical attitude, a favoring of operationally defined constructs over hypotheses, and a strong reliance on a clear-cut principle by which to bring about the elimination of all "unnecessary" elements from the world. This positivistic tradition, which may be looked upon as a narrow version of empiricism, goes back at least to Bacon who preferred "interpretations" to "anticipations" of nature, and to Newton who argued against the use of hypotheses in natural philosophy. Newton claimed that the true method, the method of analysis, "consists in making experiments and observations, and in drawing general conclusions from them by induction, and admitting no objection against the conclusions, but such as are taken from experiments, or other certain truths. For hypotheses are not to be regarded in experimental philosophy" (67:404). Contrasted with this, he argued, is the method of those who follow the Scholastics. These philosophers, who "assume hypotheses as first principles of their speculations, although they afterwards proceed with the greatest accuracy from those principles, may indeed form an ingenious romance, but a romance it will still be" (66:xx). The positivistic attitude appeared again in Comte, whose law of the three stages in the development of knowledge described man's progressive abandonment of theological and metaphysical hypotheses in favor of the purely descriptive (operational) approach. Following Comte, Ernst Mach and Karl Pearson took up the tradition and in both of them one finds a strong aversion to all hypotheses. Bavink has recently called the positivistic attitude "hypotheseophobia" (6:- 39). While one cannot say that Bridgman is a strong advocate of this position, there is, throughout his writings, an insistence on the need for defining scientific concepts by construction out of experimental data rather than by the crude methods of imagination and speculation.

It appears, then, that Bridgman is a good empiricist, both in his positive emphasis on experience as the source of knowledge, and in his denial of the need for supposing that either the *a priori* or the mystical has any cognitive role to play. He is empirical also in his strong emphasis on particulars, rather than universals. He does not succeed in reconciling his objectivism with his relativism, since he argues both for the acceptance of experience as it is given and for the dependence of our ideas upon the methods by which they are operationally derived; but in so doing he is merely taking over a conflict which has characterized the empirical tradition itself. Finally, he accepts the general spirit and approach of the positivists.

8. Bridgman's Change in Point of View

But while Bridgman has shown himself to be an empiricist in all of his writings, his earlier statements indicate a somewhat more narrowly conceived position than do his later ones. The operationism of his *Logic of Modern Physics* is dominantly a physical operationism—one might even say a metrical, physical operationism. Concepts, to be useful, must be defined in terms of measurement operations, and the operations are exhaustive of the meanings of the concepts since the concepts are synonymous with the operations by which they are obtained. To be sure, even here Bridgman does acknowledge the existence of "mental" operations, but this is hardly more than a bow of recognition; in terms of his illustrations and in terms of his exposition one gets the impression that mental operations, if not an undesirable substitute for physical operations, are certainly unimportant in the clarification of most physical concepts. The fact that Bridgman did suggest this extremely limited operationism is very significant. It is responsible not only for many of the criticisms of Bridgman which he himself, in view of his later writings, considers unfair, but also for many similarly restricted operational theories adopted by sociologists and psychologists. He has recently stated *(30)* that if he were to discuss the position again he would begin with the more general point of view, thus avoiding confusion.

In fairness to Bridgman, therefore, his later writing must be taken into consideration. When we bring these into the picture, and by means of them interpret some of the vague suggestions which he makes in his earlier writings, we come out with an operational theory which is quite different. The two positions may be contrasted in two ways, which reduce, fundamentally, to the same thing.

On the one hand, the "earlier" position seems to give operations an important, and even an exhaustive, role in the definition of meanings. *All* meanings should be operational, and meanings should involve nothing more than operations (*21*:116). This, he later admits, imposes unnecessary limitations on meanings. Operations are a necessary condition for meanings, not a sufficient one; meanings are "to be sought in operations" (*17*:103), though not always to be found there; hence all we can say is that unless we know the operations we do not know the meanings. His strongest statement of this view is that "the operational aspect is not by any means the only aspect of meaning" (*29*:257). This clearly permits the existence of meanings which are not operationally definable.

But on the other hand, the "earlier" position seems to restrict the term "operation" in a very narrow way. *All* operations should be physical and, if possible, metrical. This, he later admits, is misleading. There are many legitimate non-physical operations, variously called "mental," "verbal" and "paper and pencil," and these are significantly involved even in situations which would be described as predominantly "physical" (*21*:123; *28*:3; *23*:ix).

While Bridgman is certainly to be commended for having recognized these two sharply differentiated kinds of operational theory, one wishes that he had always taken pains, in a given context, to indicate clearly which theory was the focus of his discussion. Many criticisms which are valid against the narrow operationism are completely invalid against the broader one. Furthermore, the distinction between the two theories could have been greatly sharpened if Bridgman had provided us with a more satisfactory analysis of non-physical operations. For example, it is difficult at times to tell just what his attitude is toward verbal operations. They seem to be permitted, since verbal concepts are often helpful in producing desired action in others and may even be useful to ourselves as intermediate terms in our thinking; nevertheless the use of such methods of definition appears not to be encouraged. Most definitions of this kind resort either to an infinite regress of other terms or to terms with mystical or supernatural import. "Particularly in philosophy and religion have our verbalisms run away with us like wildfire" (20:90). In fact, the situation in these fields is so bad that Bridgman recommends scrapping all traditional philosophy and religion and starting again from the beginning. If Bridgman had provided us with a more clear-cut conception of a verbal operation, and of the way in which it is to be distinguished from other mental operations we should be in a better position to evaluate this wholesale condemnation of philosophy and religion. An attempt will be made in Chapter V to bring together various of Bridgman's statements relative to the nature and role of mental operations in thinking.

While one cannot say that Bridgman breaks with the empirical tradition in introducing mental operations, since these kinds of operation played an indispensable role in Locke's theory, nevertheless one can say that in adopting a highly generalized notion of operation he is identifying himself more closely with the pragmatists. We shall therefore turn to an exposition of the pragmatic aspects of Bridgman's operationism.

Chapter III

BRIDGMAN'S PRAGMATISM

1. The General Nature of Pragmatism

Pragmatism is much more recent in origin than empiricism. But it has, again, a variety of manifestations. For the purposes of this discussion emphasis will be placed on two of its elements. The first is the demand for clarity in ideas, and for a principle by which concepts may be made more precise. The second is the identification of truth with "workability," or, in a broad sense, the definition of truth in terms of verificatory acts rather than in terms of an absolute relation of "correspondence" holding between ideas and facts. An attempt will be made to show how these two pragmatic influences have expressed themselves in Bridgman's position.

2. Emphasis on Clarity

The emphasis on clarity has not been peculiar to pragmatism; empiricism and, indeed, rationalism have argued that obscure ideas should be reduced to a minimum, if not entirely eliminated from the knowing enterprise. But the clarity which the rationalists have sought was to be found in innate ideas, or in a special faculty called "reason," which is directed not upon particulars but upon universals. Empiricism, on the contrary, has argued that the ultimate reference of ideas, even of those which are general, must be to particulars; but since the purpose of such reference is to remove vagueness and obscurity, there is a strong emphasis on clarity. Pragmatism has continued the fight for clarity, but has shifted its emphasis from the bare particulars to the broader behavioral situation in which ideas are used. It has argued that thinking itself is simply a method by which the knower adjusts himself to his environment, and in this complex situation ideas are tools or instruments whose effectiveness is to be measured by their ability to bring about the desired end. Ideas must refer to experience, but experience itself is an interaction of the knower with his surroundings; hence an idea is clear to the extent to which it furthers smooth adjustment.

Rightly or wrongly, but by common consent, C. S. Peirce is usually credited with the origination of the pragmatic criterion for clarity. "Consider what effects, that might conceivably have practical bearing, we conceive the object of our conception to have. Then, our conception of these effects is the whole of our conception of the object" (69:5.402). "I only desire to point out how impossible it is that we should have an idea in our minds which relates to anything but conceived sensible effects of things; and if we fancy that we have any other we deceive ourselves, and mistake a mere sensation accompanying the thought for a part of the thought itself. It is absurd to say that thought has any meaning unrelated to its only function" (69:5.401). This principle was later taken over by James, who pointed out that two ideas which appear to have exactly the same consequences cannot be two but must be one.

Bridgman would presumably accept this reduction of meanings to sensible effects, and in doing so would identify himself with the entire sensationalistic tradition, both empirical and pragmatic, which culminates in the physical-

ism of some of the recent logical positivists. But there
is an important difference between him and the sensationalistic empiricists. The sensed particulars, to which
reference must be made in all clarification, are operations,
not things; if concepts are "synonymous" with operations
there can be no court of appeal but operations. This requires us to assume, in all strictness, that we can never
have the "same" construct defined by two differing operations. On the basis of Peirce's criterion, if the operations
are sensibly distinguishable the concepts must be
distinct. "If we have more than one set of operations, we
have more than one concept, and strictly there should be
a separate name to correspond to each different set of
operations" (*14*:10). To be sure, we may succeed in
showing later that these operations are within a certain
margin of error equivalent. But in order to do this we
must again employ an operational test, for there is no
method other than the use of operations by which we can
show that operations are alike or different. Thus we use
operations to clarify all concepts, and when these operations have themselves become conceptualized we use further operations to clarify *them*.

3. Emphasis on Workability

The other pragmatic factor in Bridgman is the emphasis
on activity, and, in particular, the strong reliance on workability as opposed to correspondence in the testing of
ideas. In this respect he has much in common with Dewey,
from whose writings the following statements might well
have been taken: "Science, language, rational thought,
are devices by which I try to make adjustments and I have
to find by experiment whether these are successful devices" (*19*:16). "Because of the fundamental properties of

activity itself, truth can have no such static, absolute, meaning as we would like to give it" (*19:42*). "How shall we tell in any particular case that [an analysis of experience] is good enough for our purpose? I have never heard any satisfactory answer to this question which I have not found to boil down eventually to: 'Try and see.' If a particular concept works in the way that we want it to, then it is good 'enough,' but whether it will be good enough or not cannot be told until we have tried" (*20:55*). Mathematics "is a special sort of intellectual tool, of great utility in meeting the situations of practice" (*19:117*). "The ultimately important thing about any theory is what it actually does, not what it says it does or what its author thinks it does, for these are often very different things indeed" (*19:5*). In all these quotations the pragmatic element is clearly present. While there is some evidence, as we shall see immediately, that Bridgman also favors a correspondence theory of truth, the pragmatic emphasis in these quotations is too strong to be disregarded. The entire operational method appears, in fact, to be simply a device employed by the physicist to get himself out of the difficulties which recent developments in the field have originated. Whether these difficulties are "intellectual" or "practical" makes as little difference for Bridgman as it does for Dewey. The fact that the operational theory enables the physicist to extricate himself from them is a sufficient guarantee of its truth; it works and that is all that can be required.

4. Bridgman's Theory of Truth

When we ask specifically what Bridgman's theory of truth is, the answer is not so clear-cut as one might wish. Certainly the pragmatic emphasis is strong. But some kind of correspondence seems also to be involved. Let us attempt to disentangle these two factors.

The continued emphasis on activity signalizes a dominant pragmatic element. The *true* meaning of a term, he insists, is to be found by observing what a man does with the term, not by what he says about it (*14:7*). He even contends that our understanding of his own operational point of view should be determined not so much by what he says about it as by what he does with it (*21:117*). In any case, he himself is never sure of a meaning until he has analyzed what he does (*19:9*). Furthermore, the only justification for the operational point of view itself is its utility; we have simply found from experience that if we want to do certain things with our concepts we must define them in certain ways (*21:119*); "by exposing the nature of the underlying operations, we discover whether our terms are really good for what we thought they were good for" (*86:247*). Again, truth must always be relative to operations performed by human beings in time; it can therefore have no static, absolute meaning (*19:42*). Science itself, and in fact language and rational thought in general, are simply attempts on the part of man to fit successfully into his environment; he wishes to adjust himself both to the past, in understanding what has gone before, and to the future, in enabling himself to make predictions. Whether these elaborate devices have proved successful can only be shown by experiment (*19:16-7*). Even "existence" can be defined pragmatically, for the only test of the existence of any constructions is the fact that the concepts work in the way I want them to (*19:51*).

This seems like such a clear-cut statement of pragmatism that no further evidence seems called for. But two illustrations of his use of the method indicate even more strongly the pragmatic element in his position. Suppose, for example, as is often the case in mathematics, that one is confronted with a problem whose solution is not known to exist. If he finds the solution he may comfort himself by saying that it was "really there" all the time, just as he does when he finds a lost article. But suppose he does not find the solution. Is the solution then "really there," and, if so, what can be meant by this assertion? Bridgman believes that at least part of the meaning of "There *is* a solution whether I can find it or not" lies in the *program* of the attempted solution; believing that there is a solution I shall go ahead and attempt to derive it. The adoption of a program of continued search is one which has in the past proved useful, and as a consequence has survival value (*20*:51-2). Then the "real" may sometimes become simply "an abbreviated expression of one's resolution not to give up." (*20*:53).

Consider also Bridgman's interpretation of general propositions obtained by induction. To determine what we mean by such a proposition we must examine what we do with it. "It is a rule of procedure which we accept as a valid guide for conduct in cases beyond our present experience" (*19*:34). This is, in fact, the *meaning* of induction. We find that we can formulate rules, simple and compact in form, which refer to cases of our remembered experiences. We now apply these as rules for our future behavior. The belief that these rules are eternal, have a metaphysical basis in the universe itself and are therefore a secure basis for conduct, has no operational foundation. Such principles may be used as a basis for inference, but are no more valid than the inductive process itself. A principle, in fact, may be called "general" if it is used as a major premise in a syllogism and gives us the desired *confidence* that any conclusions will turn out to be applicable in practice (*20:68*).

But while it is true that Bridgman allies himself unmistakably with the pragmatic position in admitting that meanings are operational, he goes far beyond pragmatism, as usually understood, when he generalizes the notion of operation to the point where it becomes identical with "any conscious activity" (*86:246*). We have seen how this occurred in Bridgman's thinking. Beginning with a narrow physical operationism he progressively generalized the notion to include various mental operations as well. If he had stopped with the conception of operation as any act involved in the attempt "to reduce events to understandability" (*21:115*), one might still class him among the pragmatists (though "understandability" would have to be properly interpreted); at least the *cognitive* role of operations would not have been lost. One might say, in fact, that such a general operationism is probably what Pratt meant when he said that operationism is a statement of nothing more than "the manner in which the present generation utters the familiar cry of science, 'Be careful' " (*71:81*). But when Bridgman identifies an operation with any conscious activity or, in a very recent formulation, with "any consciously directed and repeatable activity" (*48:613*), he reduces operationism, as he himself clearly sees (*21:117; 27:253*), to a tautology. For if to say that a concept is defined operationally is to say merely that it is defined in the context of activity *all* concepts become operational. Certainly all concepts take on meaning by virtue of "what we do with them"; even Newton's concept of absolute time is meaningful in the sense that it can be defined by verbal operations. Bridgman quotes with approval the conception that "science is sciencing"

(26:50). This is unquestionably true, since science is an activity carried on by human beings. But if the notion of operation is generalized to include all activity, operationism loses its distinctiveness and we no longer say anything significant when we say that science is operational or that concepts should be operationally defined.

Furthermore, this pragmatic emphasis on activity is not the whole story, for we find in Bridgman also a strong inclination toward some sort of correspondence theory of truth. He says, in fact, that the success which we achieve in our use of language must be due to "its ability to set up and maintain certain correspondences with experience" (19:19). He then goes on, and in the mood of a pragmatic Wittgenstein argues that *why* we are able to set up such correspondences and to maintain them is a meaningless question; we can simply accept the fact that we do so. "The processes of establishing the correspondence through which language has meaning cannot themselves be described in language, but involve getting outside the system of language" (19:20). This correspondence is not, of course, close or detailed. In general the structure of a language and that of experience are not the same. For, in the first place, experience is flowing, changing and reforming; language has only discrete units held together by remembered connections. Furthermore, language does not provide a unique method for reporting isolated aspects of our experience. Finally, the lack of correspondence is clearly demonstrated by the fact that all sorts of linguistic structures having no known correspondence with any experience are perfectly possible (19:21-3).

The best illustration of the kind of correspondence which does hold is to be found in mathematics. Although Bridgman's view of mathematics, as we have seen (p. 18), is not unambiguous, he does argue that mathematics is a human invention. Furthermore, since it has been created by the physicist for the purpose of describing the external world there is no accident in the fact that a correspondence holds between mathematics and nature. But this is far from perfect, for two reasons. In the first place, mathematics applies only approximately, due to the errors of measurement (*14*:62-3). In the second place, and more importantly, mathematics tends to encourage an incautious empiricism, for it allows its equations to extend beyond the areas where they have strict application (*19*:-68). For example, so far as the equations of motion are concerned there is no principle which prevents them from being extended to the motions of stars in our galaxy on the one hand, or to the motions of an electron about the nucleus on the other, even though the physical meanings of the quantities are quite different in the two cases. Mathematics, in fact, reminds one of the loquacious orator who is able to "set his mouth going and then go off and leave it" (*14*:63). It tends to encourage the pre-operational attitude toward concepts, by which they are presumed to have application in all areas, in spite of the fact that their origins lie in highly restricted regions. Hence mathematics, even if apparently descriptive of the world, contains reference only to a very small part of the actual physical situation. Included in this broader background are all the physical operations by which the mathematical data are obtained (*14*:60-5). What mathematics requires, therefore, is a "text" (*19*:59). For example, in Einstein's theory the data are merely numbers, *viz.*, space and time coordinates. To make contact with experience these should be placed in a descriptive background giving the physical contents of the events, *e.g.*, that some of the events are light signals. Apart from this background mathematics is too abstract to serve as a description of nature.

With regard to the question, therefore, whether Bridgman advocates the pragmatic theory of truth or the correspondence theory, no clear-cut answer is possible. On the one hand, he seems to rest utility on correspondence; concepts work because they correspond. But, on the other hand, the only justification for the operational point of view itself is its utility; we have simply found from experience that if we want to do certain things with our concepts we must define them in certain ways (21:119). Thus by exposing the nature of the underlying operations, we are able to test whether our terms will really do what we thought they would do.

Chapter IV

OPERATIONISM IN THE SCIENCES

1. Variations in the Scope of Operationism

AMONG SCIENTISTS operationism has called forth widespread and varying reactions. In general, those who have discussed the method are divided into the protagonists and the antagonists, but the line of demarkation between the two groups is far from clear. The reason for this is apparent. As indicated in the case of Bridgman, the word tends to take on greater and greater generality during the course of discussion. As a consequence of this shift in the meaning of the term, operationism becomes readily identifiable with a wide range of positions extending from the narrow insistence that only quantitative methods may be used in science; through the vague demand that science be empirical, as opposed to rationalistic, or metaphysical, or speculative, or any other notion with which the term "empirical" is commonly contrasted; and ending with the extremely general position that any concept is operationally defined if it produces "understanding." One can readily see that the two extreme positions are not by any means identifiable with one another, and that neither of these is strictly identifiable with the more moderate position that lies between them. Since our purpose in this chapter is to emphasize some aspects of operationism rather than to characterize certain writers as definitely "for" or "against" the position, this variation in meaning will present no serious problem.

(42)

2. Operationism in Physics

In view of the great variety of ways in which Bridgman has used the term "operation" one is not surprised to find that both among those who have followed his lead and among those who have rejected his proposal there is essential disagreement as to what the term, in the final analysis, means. The physicists have quite generally either accepted Bridgman's position or, strangely enough, more or less completely disregarded it. The disregard, one would suppose, implies an acceptance of his position; if, as Bridgman maintains, he is merely describing the usual method of physics, there is no reason why any physicist should object to what he is saying. Philipp Frank (*45*:4-5; *46*:44-5) describes Bridgman's operationism with apparent approval. Lenzen (*51*:367) states that the point of view "now appears to be generally accepted in physics." He considers Bridgman's operational definitions to be statements of the relations between pure numbers and physical objects (*52*:30)—an interpretation which Bridgman would, I am sure, be quite willing to accept, at least so far as metrical definitions are concerned. Margenau incorporates Bridgman's operational definition in the more general concept of *epistemic definition*, which is contrasted with *constitutive definition* (*62*:232 *et seq.*). Northrup has a similar distinction between *concepts by intuition*, which are defined by epistemic correlations (operations), and *concepts by postulation* (*68*:ch. 5). Lindsay is one of the few who have criticized Bridgman's operationism. The grounds for this rejection are, surprisingly enough, that Bridgman has not in fact described the usual method of physics; Lindsay argues that the adoption of operationism would, on the contrary, make all theoretical physics impossible, and that the success already achieved by the existing method should make physicists cautious about sacrificing it on the altar of operationism (53:470). In the joint study by Lindsay and Margenau (54:412) the objection is raised that definition in terms of measurement is not always feasible.

3. Operationism of S. C. Dodd

One of the leading sociologists who has defined operationism in terms which would seem general enough to eliminate all anti-operationists is Stuart C. Dodd. He states that a definition is operational "to the extent that the definer (a) specifies the procedure . . . (including materials used) for identifying or generating the definiendum, and (b) finds high reliability . . . for his definition" (39:482). A "procedure" is defined as any human action which is a means to an end, and communicable by the actor. "Reliability" is defined as "any index . . . measuring the degree of agreement . . . among reobservations of the same phenomenon"; this means that "a concept is reliable in proportion as the designata are constant for all interpretants under specified conditions." All of this reduces, he suggests, to saying that "an operational definition is any statement . . . which reliably tells what to do, first, second, third, and with what ingredients, in order to test for the presence of, or to produce, that which is defined" (39:484).

Several features of this definition should be called to attention. In the first place, there is no specification that the operations be metrical, or even that they be physical. He explicitly states that operational definitions are not limited to quantitative ones, since appropriate formulae exist for determining the reliability of qualities as well. Thus any process by which the referent of a term may be reliably identified will suffice. Presumably this means that verbal operations, and even imaginative operations, so far as they can be communicated, are legitimate types.

In the second place, it should be noted that Dodd makes explicit reference not only to procedures but also to materials used. In this respect he would almost certainly disagree with Bridgman's statement that a concept is *synonymous* with certain operations. Dodd seems to realize clearly that merely indicating procedures (operations, in the narrow sense) would leave out the substantival element which is part of the referent of every concept; operations must always be performed on *something*.

In the third place, Dodd attempts to clear up an ambiguity in Bridgman's formulation by distinguishing between *identifying* and *generating* the referent of a concept. We may specify the referent of the word "cake," for example, either by finding one or by baking one. Bridgman's position is not clear in this regard; he suggests at times that lengths are *created* by measurement, and at other times that they are merely *identified* or recognized by means of the measuring activities. Dodd has certainly improved upon Bridgman in explicitly recognizing this distinction; there is an advantage in knowing that in order to identify a cake we need a good definition, but in order to create one we need only materials and a recipe. However, even in Dodd's improved formulation an ambiguity still remains. He seems to allow for the possibility that the *mere baking* of a cake could be operational without any supplementary definitional operations; but this seems incompatible with the almost universally recognized principle that operationism is a device for creating and defining *concepts*, not for producing *things*.

Finally, Dodd replaces the demand that operations be public and repeatable by the simple requirement that the definitions which employ them be reliable. Now reliability can itself be measured. Thus definitions are not operational or non-operational; they are only more reliable or less reliable, and *any* definition, regardless of how it is formulated, will be acceptable provided it has a high degree of reliability. To the extent, therefore, that all scientists, operationists and non-operationists alike, accept as one of the ideals of science *precisely defined concepts*, and to the extent that they are willing to employ *reliability* as a measure of this precision, none could object to the use of operational definitions. Hence Dodd is an excellent example of an operationist who defines his point of view in terms so general that few can disagree with him.

4. Operationism of G. A. Lundberg

A somewhat more narrowly conceived operationism has been defended by G. A. Lundberg. His contributions to the theory illustrate a point which is highly significant in the consideration of the logical foundations of operationism, *viz.*, that to the extent to which one insists on a narrow interpretation of the term "operation" one is compelled to concede the insufficiency of the operational method in science. Although Lundberg begins with a very general conception of operation, he ends with a behaviorism. The main reason for introducing operational definitions, he argues, is to avoid "ambiguousness and lack of precision in the definition of concepts" (58:728); these are *never* desirable in scientific work. In the most general sense "there is nothing mysterious or magical about the word 'operational'. It is merely a word we use to designate a type of communication which 'gets across' with high reliability. Thus, a recipe for a chocolate cake may be regarded as an *operational definition* of such a cake" (60:89). Objective data, used inductively, must be the basis of science (56:394). However, the objectivity which we seek is "regarded not as a character of things but as those ways of responding which can be corroborated by others" (56:192). Hence, in spite of the fact that "actually the simplest form of operational definition of a word is to point to its referent while enunciating the word" (58:730), we cannot know any type of data except through the operation of symbolic behavior mechanisms, exhibited in the form of language (56:397).

For Lundberg, therefore, the demand for objectivity in science is satisfied by operational definitions. But objectivity, instead of being a property of things, is really a property of our ways of responding to things, particularly of our symbolic behavior mechanisms. Consequently, in order to produce highly reliable knowledge of the world we have only to pay particular attention to our linguistic behavior. One can readily see that Lundberg's behaviorism is more general than others which have been proposed, since he includes all symbolic processes under the category of behavior. But the strong operational slant is clearly indicated in his insistence that in the definition of a concept the reaction of some organism to the object presumably designated by that concept be included.

Apparently, however, Lundberg recognizes the impropriety of a rigid operationism, and grants that in spite of the fact that "all statements about the nature of the universe or any part of it are necessarily a verbalization of somebody's responses" we must admit something which can be called "*that which* evoked these responses." "The nature of that which evoked them must always be an inference from the immediate datum, namely, our symbolized sensory experiences." But then having made this concession he immediately sets about to protect himself by insisting that "all assertions about the *ultimate* 'reality,' 'nature,' 'essence' or 'being' of 'things' or 'objects' are . . . unverifiable hypotheses, and hence outside the sphere of science" (59:9). In the definition of every concept there is, therefore, a non-operational element—an element, moreover, which must remain forever non-operational since it is definable only as that which evoked certain responses. And no matter how we may multiply responses in the future we shall never be able to get at this element operationally.

Closely related to this concession is the reply which Lundberg makes to the criticism that operationism limits the meanings of concepts. He readily grants that this is the case, but that it should not disturb us since *all* definition limits meaning. But he concedes further that in insisting upon the operational definition of a word operationists do not "propose to deny or ignore all other referents which this word in its traditional meanings may have had" (58:734). He even grants that some scientists may be primarily interested in those stages in the development of science where "intuitive methods must necessarily play a large part, as compared with the partly charted areas in which refinements of methods are the chief conditions of advance." But then he seems to feel that he has conceded too much for he insists that the "intuitive procedures should be supplanted as fast as possible by the more objective ones" (58:738), or, as he expresses it in another context, "these *other* meanings, in so far as they are relevant to the problem we have set ourselves, must be similarly defined operationally" (60:92-3). But if we grant the necessity for introducing the notion of "that which evokes responses," which we can never even hope to define operationally, we seem committed to the admission that there are some "meanings" which are known only by "intuitive procedures." Further reference will be made to this important aspect of operationism in the ensuing chapters.

5. Operationism of F. S. Chapin

A still more narrowly conceived operationism is defended by F. Stuart Chapin (*35, 36*). According to him a definition may be one of three kinds: verbal, heuristic or operational. The verbal definition is the traditional kind; for example, "morale" may be defined as "the degree to which the individual feels competent to cope with the future and achieve his desired goals." The heuristic definition "uses diagrams, figures and graphic forms, to represent the concept as an entity" (*36*:158); it serves mainly for temporary and transitional use between the verbal definition and the more useful operational definition. "Advance in research that is sound should seek to replace verbal or word definitions of social concepts with operational definitions as rapidly as research permits" (*36*:155). The operational definition "depends upon the series of acts performed by the investigator in the process of measurement . . . Scales to measure public opinion are constructed and tested for validity. We say of a scale that has been standardized to measure public opinion, 'Public opinion *is* what this scale measures.' This is then the operational definition of the concept public opinion" (*36*:156). Such a statement is susceptible of check, test or verification. In other words, it is a more objective definition than either the verbal or the heuristic type.

Here Lundberg's behaviorism has taken on a more specific character, for objectivity as conceived by Chapin lies not in *all* "ways of responding" but only in those which are metrical in character and eventuate in number scales. The manner in which this objectivity can be achieved may be illustrated by Chapin's discussion of the concept of socio-economic status (35:Ch. XIX). Let us start with a verbal definition of "socio-economic status" as the position that an individual or a family occupies with reference to the prevailing average standards of such things as cultural possessions, effective income, material possessions, and participation in the group activities of the community (35:374). The problem is to objectify this concept. One method which suggests itself is to examine the living room of the home of each family whose socio-economic status we are to determine, and to evaluate what we find there. This may be done by noting the number and kinds of fixed features (floor, floor coverings, windows, *etc.*), of built-in features (bookcases, beds, window seats, *etc.*), of items of standard furniture (tables, chairs, *etc.*), and of cultural resources (pictures, books, newspapers, telephone, *etc.*) which are present. Values, variously weighted, can be attached to these items, and a number can be computed which may be taken as the measure of the socio-economic status of the family.

Such a score must first be tested for reliability. This is done by having various visitors examine the same homes, and by having the same visitors re-examine homes which they have previously measured. If the correlations are sufficiently high the test may be considered reliable. This establishes the fact that the test measures consistently *whatever* it is designed to measure.

But at this point one is tempted to raise an objection.

How, it may be asked, can we be sure that the test *really* measures what we intend by the term "socio-economic status"? We can adopt an operational definition and say that socio-economic status *is* just what the scale measures. But this does not seem to meet the problem. Unquestionably the number obtained by this method of surveying the living room measures *something*. But does it measure socio-economic status?

Chapin grants that this objection is apparently destructive of the very basis of operationism (36:156). But that this is only apparently so is readily seen when we recognize that socio-economic status and its measured value should not be identified until *after* the scale has been standardized. To examine a test for *reliability* is not sufficient; it must also be examined for *validity*. "The process of standardization, if done thoroughly, disposes of the question of validity, so that the assertion of the operational form of definition . . . does not beg the question" (36:156). The test for validity as applied to the measurement of socio-economic status requires us to examine other scales designed to measure the same thing, and to ascertain the degree to which they are consistent with one another. Income suggests itself immediately as another way of measuring socio-economic status, and one finds that the correlation between the income scores and the living room scores is significantly high. Similar correlations may be obtained between living room scores, on the one hand, and scores measuring occupation, participation in community affairs, and education of wife and husband, on the other. The presence of these high correlations guarantees that what we are measuring is *really* socio-economic status, and the term may now be defined operationally in terms of the "combination" or "configuration" of these special scores. Socio-economic status is a "pattern of parts forming a fluctuating whole" which then functions as a unit stimulus calling forth a response (*35:-373*).

The significant contribution of Chapin, therefore, is not that he defines operationism in terms of metrical procedures (Bridgman is generally considered also to have favored this interpretation), but that he allows complicated functions of more directly derived measured values to have the same "objective" status as the measured values themselves. Socio-economic status is not "directly" quantifiable but is measured only by means of complicated coefficients of correlation between other more immediately obtainable quantities. The question which naturally arises is whether *all* mathematical functions of directly derived quantities, or only *certain* of these functions, will have "objective" status. Clearly one has the privilege of performing any mathematical operation he pleases on empirically given quantities; furthermore, his operations may be carefully specified and then will be sufficiently objective and public to satisfy any except the most narrowminded operationist. But will he be measuring anything? Obviously he cannot be sure that he will be measuring what he set out to measure, for this is not given as a directly measurable quantity. And if not, what will he be measuring? This question will be further examined in Chapter V.

6. Operationism of S. S. Stevens

Psychologists have become increasingly interested in operationism in recent years (*2, 3*). Among those who have been strongest in their advocacy of the operational method is S. S. Stevens. As in the case of almost all writers on operationism he gives due credit to Bridgman, whom he describes as "simple and hard headed" and as one who, in spite of the fundamental soundness of his position, must expect criticism because of the vital issues which he challenges. Furthermore, Stevens asserts, since psychology is more difficult than physics there is bound to be more dissension among psychologists than among physicists as to how operational principles are to be interpreted and applied.

Yet the term "operationism" remains, even after Stevens' discussion of it, extremely vague and general. It represents the "empirical" point of view (63:27), yet we are given no information as to how this word is to be interpreted except that operationism has great value in eliminating even the "choicest propositions of metaphysics" (63:23; 83:527). The essential demands of an operational theory are that the operations be "public and repeatable," and that they be "concrete" (63:27, 28); in this respect he regards Bridgman's admission that there may be "imaginary" operations (17:110) as conceding too much (82:-324). Fundamental among operations is that of "discrimination" or "differential response," and the sole business of psychology is to test and measure these discriminatory capacities of the organism (82:325). This suggests behaviorism, but Stevens denies the identity of the two positions. Behaviorism erred in denying too much; "operationism does not deny images . . . but asks: What is the operational definition of the term 'image'?" Furthermore, operationism is not to be identified with the traditional positivism, which "pretended to base *everything* on experience" (63:30), and thus failed to allow for the formal methods of mathematics and logic. What Stevens' operationism reduces to, therefore, is a vaguely conceived empiricism in which "the empirical" is to be defined in terms of the public, the repeatable and the concrete, and to be contrasted with both the metaphysical, which it supplants, and the formal, which it employs as a methodological tool.

7. The "Symposium on Operationism"

Equally unsatisfactory, from the point of view of any fundamental clarification of the term "operationism," is the "Symposium on Operationism" *(86)*. Of the contributors only one, Israel, has raised significant objections to the general viewpoint, and his criticisms are not answered by any of the other participants. Feigl directs his discussion more or less to the heart of the matter and ends with a summarizing statement of qualities which acceptable operations in the factual sciences should possess. They should be logically consistent; sufficiently definite (if possible, quantitatively precise); empirically rooted, *i.e.*, by procedural and finally, ostensive links with the observable; naturally and, preferably, technically possible; intersubjective and repeatable; and aimed at the creation of concepts which will function in laws or theories of greater predictiveness (*86:258*). He then suggests a classification of kinds of operational definitions into purely qualitative, semi-qualitative ("comparative" or "topological") . . . fully quantitative or metrical, causal-genetic, and theoretical constructs. Pratt calls attention to a distinction between two stages in the development of an hypothesis; "in the first stage the observations, operations, and correlations of an experiment are summarized by some phrase which can be given operational definition in terms of what was observed and done in an experiment, or in clinical practice." In the second stage the definitions are rephrased so as to intend "hypothetical agents which are thought of as having a real existence. This second stage is likely to be guesswork and imagination, and is therefore objectionable to some operationists and pure correlationists . . . The further the scientific imagination goes beyond the safe circularity of the first stage of hypothesis-construction, the more improbable it becomes that the agent postulated in the hypothesis can be given operational definition" (*86:267-8*). This is an important distinction, and will be elaborated later in the views of certain other psychologists. Skinner defines operationism in terms of observations, manipulative and calculational procedures involved in making them, logical and mathematical steps intervening between earlier and later statements, "and *nothing else*" (*86:270*). This emphasizes the negativistic aspect of operationism. But in including in his conception of operation all logical and mathematical steps, *i.e.*, presumably all *inferences* which may be made from observations and measurements, he generalizes the term in a manner which Feigl (*86:252*) considers undesirable. Furthermore, one can readily see that to use the term in this highly general sense would immediately make all scientists operationists, since one could hardly conceive of any entity entering into scientific consideration which is not either observational or "inferred" from something observational.

Certain other psychologists, in attempting to clarify the operational point of view, have taken over and renamed a distinction (*8:183-9*) which has played an important role in the philosophy of science since the famous "*Hypotheses non fingo*" of Newton. Because the distinction is of great importance for the clarification of the limits of operationism the problem will be considered in some detail.

8. Descriptive and Postulational Concepts

It is really a two-fold problem. In the first place, a distinction is required between concepts which are more or less directly descriptive, on the one hand; and concepts which involve the use of methods of construction, inference, and insight or creative imagination, on the other. The need for this distinction arises as soon as we recognize that science is not confined to the mere listing and describing of data, but involves exploratory and inventive activities which are designed to anticipate what nature will later disclose to direct awareness. Northrup (68:82-3) calls the symbols which are employed in the former task "concepts by intuition," and those which are used in the latter "concepts by postulation." Bridgman (14:53-60) calls the latter "constructs." Feigl (44) distinguishes between the language of data and the language of constructs. Margenau (62:44-6; 61) employs a similar terminology and claims that constructs are a special type of "auxiliary concept" which "release the scientist from the bondage of sensory experience." Since the limits of what can be strictly described are not precisely determined, the distinction between descriptive concepts and constructs is not a precise one; there will almost certainly be concepts which cannot be unequivocally placed in either class. But the distinction is a convenient one from the terminological point of view. For example, the concept of color can be considered to be descriptive, and the concept of the atom, by contrast, constructed or postulational.

9. Constructed and Inferred Concepts

But once this distinction has been granted, another arises *within* the field of concepts by postulation. Here the terminology becomes very much confused. As we have just seen Margenau calls all such concepts "constructs." Russell (76:155-8) subdivides them into two types, "constructions" and "inferences"; Reichenbach (73:211-2) distinguishes between "abstracta" and "illata" and contrasts both of these with impressions, or "concreta," which presumably are descriptive concepts; Bavink distinguishes between "elaborative" and "hypothetical" theories (6:34); and Cassirer (34:193-4), Dingle (37:22-4) and Hobson (50: Ch. II) contrast the method of "abstraction" with the method which produces "hypotheses."

It would be rash to assume that all of these writers have exactly the same distinction in mind. Yet there is probably a basic contrast which they are trying to express. We seem to be able to "go beyond" data by two fundamentally different methods; we can *construct* (invent) new concepts, or we can *infer* them. In both cases the operations have their origin in certain data or intuited concepts; and in both cases the operations may be more or less unconsciously performed, thus creating the impression that the constructed and inferred concepts are *discovered* rather than *derived*. But in other ways the methods are sharply contrasted. Construction produces concepts with merely a "linguistic status"; they refer to nothing directly and refer indirectly only to the data or intuited concepts which originated them. In this way the number and variety of objects in the world is kept at a minimum, since the diversity of symbols indicates merely a diversity in modes of representation, not a diversity of things referred to. The method is significantly employed by the nominalists and the conceptualists, both of whom interpret *universality* as an indirect manner of referring to particulars rather than as a direct mode of referring to universals—

an interpretation which is in accordance with Occam's razor, since the world is thereby "rid" of a certain type of entity. In much the same way the concept of the average man may be considered to be a linguistic construction out of actual men, and the concept of the perfect gas to be a linguistic construction out of actual gases.

Inference, on the other hand, is something quite different. While inferred concepts also have their origin in certain data or intuited concepts, to which they then refer indirectly, they are also assumed to refer directly to certain elements of the world having the status of potential data. By inference we extend not merely our symbolic scheme but the realm of the given. Universals, for the realist, are natural elements which can be discovered in the world. Similarly, atoms are commonly considered to be inferred rather than constructed concepts; atoms are presumed to exist in some sense. While constructed concepts are given content through operations of "free creation," inferred concepts must be restricted to what we expect later experience to disclose. In the words of Bavink, inferred concepts involve the "presumption of the existence of a general state of affairs lying at the back of certain phenomena which are matters of experience" (6:42). This reference to existence is, of course, purely tentative until confirmation has been completed; but inferred concepts are at least assumed to have existential reference, while constructed concepts are not.

The importance of this distinction in the formulation of operationism can now be more readily seen. Operationally, constructed concepts are much to be preferred to inferred concepts. While for a strict operationist such concepts would be forbidden (since the operations involved in their creation are not physical in character), for the more liberal-minded operationist constructs are perfectly legitimate. They have two important virtues as compared with inferred concepts. In the first place, they are, in general, more clearly defined than inferred concepts. Although, as we have just seen, both constructed and inferred concepts may be derived unconsciously, nevertheless unconscious inference is somewhat more likely to occur than unconscious construction; when we infer we are frequently unaware both of the fact and of the manner of our doing so, but when we construct we do so more or less deliberately and with a fairly clear-cut awareness of what we are doing. This is not to deny either explicit inference or unconscious fabrication. But it means that sharp definition of constructed concepts is somewhat easier to achieve than of inferred concepts. Constructed concepts are much like descriptive concepts in that, ideally at least, all of the factors that enter into their meanings can be accurately specified; just as the concept *redness* means certain actual reds, as operated upon by generalization, so the concept *perfect gas* means certain actual gases arranged in order of "perfection," as operated upon both by generalization and by serial extrapolation to a limiting case.

But constructed concepts have another advantage over inferred concepts which makes them even more to be preferred on operational grounds. Constructed concepts involve no existential commitment, since they are merely linguistic conveniences. Inferred concepts, on the other hand, require such commitment. If one shares, as so many operationists do, the positivistic point of view that science should concern itself with certainties, and should rarely if ever be venturesome, he tends strongly to favor constructed concepts over those which are inferred.

Much of the discussion of the role of operations in psychology centers about the problems of whether, among postulational concepts in this field, constructs suffice or must be supplemented by inferred concepts. As will be seen immediately, the lack of an agreed-upon terminology is a source of great confusion.

10. Role of Intervening Variables in Psychology

E. C. Tolman is responsible for introducing the notion of "intervening variable" (63:87-102). This is necessary, he feels, in order to account for the role which mental processes, either of ourselves or of another, play in the prediction of behavior. Mental processes, so far as their "cash value" is concerned, are simply a set of intermediary functional processes which connect the initiating causes of behavior, on the one hand, with the final resulting behavior itself, on the other (63:88). Mental processes are but intervening variables between the five independent variables of environmental stimulus (S), physiological drive (P), heredity (H), previous training (T) and maturity (A), and the dependent variable, behavior (B). On the basis of experimentation we set up the tentative formula

$$B = f_1 \ (S,P,H,T,A)$$

Since this is a very complex function, whose nature is completely unknown except in the simple cases of reflexes and tropisms, we have to break it down into component functions. We do this by means of "logically constructed intervening variables" (I_a, I_b, I_c). Now if we can show that these intervening variables are certain functions (f^a_2, f^b_2, f^c_2) of the given independent variables, then we can show that behavior itself is a certain function (f^x_2) of the intervening variables plus the independent variables. In this way we shall achieve the desired functional statement, *i.e.*, we shall now be able to correlate behavior with both its empirical conditions and certain intervening variables which are themselves defined as logical functions of these empirical conditions. These intervening variables may be considered either as "mental events" or as "mental capacities", but the point is that they need not be introspectively defined (Tolman "doubts" that one can get at them introspectively (63:100)). On the contrary they may be defined in strict accordance with operational principles.

Taking Tolman as a starting point, K. MacCorquodale and P. E. Meehl (63:103-111) refine the notion of intervening variable and introduce an opposing type of concept which they call the "hypothetical construct." The contrast is then rendered as sharp as possible by pointing out that the intervening variable has four characteristics, none of which is possessed by the hypothetical construct. It is "simply a quantity obtained by a specified manipulation of the values of empirical variables; it will involve no hypothesis as to the existence of nonobserved entities or the occurrence of unobserved processes; it will contain . . . no words which are not definable either explicitly or by reduction sentences in terms of the empirical variables; and the validity of empirical laws involving only observables will constitute both the necessary and sufficient condition for the validity of the laws involving these intervening variables" (63:105). The validity of intervening variables cannot be criticized except by denying the empirical facts; hence all that we can say of them is that they are "convenient" (63:107). Hypothetical constructs, on the other hand, since they assert the existence of unobserved events or processes, are true or false.

Melvin Marx (63:112-128) accepts this distinction, and directs his attention to two further questions: Is there a difference in function as between the two types of concepts which might justify their continued supplementary use in science, and is there in science a special type of intervening variable which might be regarded as an operationally valid alternative to the hypothetical construct? He answers both questions in the affirmative. Hypothetical constructs are needed in the "preliminary phases of a new scientific development." The advancement of science requires the asking of "straightforward questions which can be given direct empirical answers," but these questions can be derived only from some kind of prior hypothesis. Now if these hypotheses cannot themselves be operationally defined, as is usually the case in the early stages of science, we must rely on "operationally inadequate conceptualizations"; and we must think of the hypothetical constructs merely as suggestive of research, not as themselves having any empirical validity. Furthermore, they must be looked upon as merely temporary devices and to be transformed as quickly as possible into "operationally purified intervening variables" (63:117). Thus hypothetical constructions are stop-gap devices, necessary in beginning science, but having no place in "sound scientific theory" (63:127). The special type of intervening variable which Marx introduces is described by the symbol "E/C" and is designed to indicate the difference between the experimental group and the control group in an experimental situation. The two groups are, of course, so far as possible, identical except for the variation of one factor. If when the stimulus is varied in the experimental group the response is also varied, then the experiment may be considered successful. The symbol

"E/C" may be employed to indicate "nothing more than whatever intervening function needs to be assumed in order to account for the experimental-control differences empirically observed" (63:124). As can be readily seen this is a firm attempt to avoid any theoretical element and to formulate the results of the experiment in strictly operational terms.

Psychologists, therefore, have taken over operationism largely because it has seemed to many of them to provide a solution to the troublesome problem of whether they need in their science specific symbols for mental processes. On the one hand, if one thinks of such processes as hypothetical constructs (inferred concepts), he must be prepared to find his symbols vague, and, at best, incompletely confirmable; hence he might better either use them only in the beginning stages of science or dispense with them entirely. On the other hand, if he thinks of mental processes as intervening variables (constructed concepts), he seems to be able both to eat his cake and have it; such processes are not, of course, directly observable, but vagueness in talking about them is eliminated by defining the variables through "specified manipulations of empirical variables," and inconfirmability is avoided by defining them in such a way as to make empirical laws the "necessary and sufficient conditions for their validity." But if a variable is to "intervene," and yet be "empirical" without being "hypothetical," it must have the purely linguistic status of an average or an idealization. The further problem would then be whether a science which dispenses with all but such linguistic tools could ever progress.

Chapter V

CRITIQUE OF OPERATIONISM

1. Scope of the Problem

A s HAS already been indicated, Bridgman seems to have shifted in his conception of operationism from a narrow point of view in which the meaning of a concept was identified with certain operations, preferably physical in character, to an extremely general one in which the meaning of a concept was defined in terms of any use to which the concept could be put. A similarly wide range of interpretations is found among the sociologists and psychologists who have taken Bridgman's view as their point of origin and who have adopted it with or without modifications. As a result the critic of operationism is confronted with two alternatives, neither of which is completely satisfactory.

On the one hand, he might take the most narrowly conceived operationism as representative of the position. This would require him to reject all of Bridgman's writings after his *Logic of Modern Physics,* and even some "hints" in this work itself, and would oblige him to disregard all followers of Bridgman who have generalized on his conception. But the attack would then be simple; it would consist of a flat denial. No concept is, or could be, synonymous with any given set of physical operations. Operationism in this sense is both false as a description of actual techniques of knowing and utterly futile as a proposal for methods of the future. Thus the "early" Bridgman and his most rigid followers would seem to be arguing for something which is patently absurd. On the other hand, the critic might take as representative the considerably generalized conception which Bridgman presents, especially in his *Intelligent Individual and Society,* but also in many of his other later writings. Here he seems to be saying nothing more than that operations of some kind usually enter into the definitions of certain concepts. This is a point of view with which no one could possibly disagree. Margenau points out that an operationism of this kind "would not be saying much" (62:232), and Brunswik indicates that this is what happens to the theory when we do not restrict it sufficiently and allow "relatively casual testing procedures including introspection" to be placed on an equal footing with objective methods (31:13). In view of these two extreme positions the critic is confronted with the task of analyzing a position which is either clearly false or obviously true, and in neither case can he have anything important to say.

To adopt either of these attitudes, moreover, would be not only unfair to Bridgman and his followers, but would disregard completely the fact that the operational point of view has had a tremendous impact on modern scientific thinking. A theory which was clearly false could not possibly claim the large number of adherents that operationism does today, nor could one which is perfectly obvious produce the large number of critics. One suspects, therefore, that there must be an element of truth in the position—a strong element, in fact—which seems to lend itself only too readily both to overstatement and to understatement. The problem is to extract this kernel of truth, and to give it a formulation which is as concise as possible.

A third alternative might seem open to the critic. Instead of criticizing any existing conception of operationism he might point simply to the absence of any agreed-upon definition, or, what amounts to the same thing, to the large number of definitions, not equivalent to one another, which have appeared since the publication of Bridgman's first book on the subject. He could then take his stand on the thesis that until a satisfactory definition is forthcoming, *no* criticism of the position is possible. But this would only confuse the issue further by inviting another flood of definitions. These would be contributed not only by new converts to operationism, but (what is more important) by the former advocates of the position, who, like Bridgman, progressively broaden their interpretations in the light of criticisms. Operationism, in common with the closely affiliated position of logical positivism, began with a point of view which was perfectly clear-cut but obviously absurd. Then, through revisions and reformulations it achieved a greater and greater generality, with an ever increasing ambiguity. As a consequence, there is no definable meaning running through all the interpretations; the word should in all strictness always be written with a subscript, indicating both the name of the definer and the date at which he formulated his definition. But this would make the criticism of the position an almost impossible task.

In this chapter the attempt will be made to bring together some of the most common criticisms of operationism. That certain of these will be invalid for some types of operationism follows from what has just been said. Those criticisms which have been selected for consideration are deemed to be basic, and such as to apply to any except the most generally conceived operationism.

2. Over-Emphasis on Particularity

The first criticism is the tendency on the part of operationism, especially as exemplified in Bridgman, to place undue emphasis on the *particularity* of every operation. This seems to be the common complaint of a number of critics of operationism, among whom are Lindsay (53:457-8), Israel (86:261) and Benjamin (7:442-4). We saw in the discussion of the historical antecedents of the position that this stress upon uniqueness and individuality is found in the entire empirical tradition; for the empiricist the particular is ultimately more real than the universal, and therefore the final court of appeal in all matters of knowledge. Bridgman emphasizes the fact that an operation is an act performed by a given individual at a certain time and place. But if this is true another operation, however similar to the first, must be a different one since it will be distinguished at least by spatial or temporal location. Two measurements of the length of a given object, even if the results are the same, can be distinguished. Now if a concept is always to be defined by an operation, and each operation is a particular, the concept itself takes on the particularity of its mode of definition. Not only will there be a difference between the tapeline length of a field and the triangulation length (even if the measured values are the same), but there will be a difference in *meaning* between all individual tapeline lengths of the field (again, even though the measured values are the same).

Although some operationists are not so explicit in their insistence upon particularity, they adopt a position which readily lends itself to the same interpretation. The continued reference to the *observable* leaves no alternative, for only particulars can be observed in the strict sense.

To be sure, two particulars may be observed to be similar to one another, but so also, as we shall see in a moment, may two operations. Nevertheless, in seeing them as two we recognize their differences, and these differences, on a strict operational interpretation, would have to be included in any concepts which we employ to designate them. The insistence by Stevens (*63:28*) and by Hart (*47:293-7*) on concreteness presumably reduces to the same thing. While the word "concrete" is ambiguous, referring either to the particular (the individual) or to the specific (that which has a low level of generality), the context of Stevens' discussion seems to indicate that the reference here is to the *particular* rather than to the *kind*. Finally, the almost universal insistence by operationists on the need for *clarity* in our conceptual tools again suggests the reference to particulars, for all generality involves a greater or lesser degree of vagueness.

To this charge Bridgman might well reply that it neglects entirely his emphasis on repeatable operations (*86:-246*), on operations which can *always* be performed (such as the mathematical operation of adding one to an integer) (*21:124*), and on the equivalence of different operations which give "very nearly or perhaps the identical result" (*21:121*). Thus he might insist that a concept is defined not by a unique operation, performed here and now, but by a *class* of operations connected with one another either by resemblance or by some other mode of correlation. This may, in fact, be what Bridgman has in mind in speaking of a "set" of operations. This term is ambiguous since it may mean a *class* of operations, or a *series* of operations exhibiting a certain sequence. In any case we are not told to what degree operations must be similar for them to constitute a set.

But again one is confronted by the obstinate fact that Bridgman seems to feel that experience is ultimately only of particulars. This can be seen in his continued reference to *himself* and *his own operations* (*19*:8-9, 83; *20*:21), when he attempts to determine the meaning of any concept. He even suggests, as we have seen, that his operational theory is to be understood in terms of what he *does* with it, not in terms of what he *says* about it. And he has on many occasions (*19*:13-4; *26*:43-61) insisted that science is ultimately and finally a subjective matter—a matter of what the individual himself thinks and does. Furthermore, he has claimed that unless we define in terms of operations *actually performed* we run the risk of using self-contradictory or physically impossible operations (*86*:246-7; *19*:9). Still again, operations should not be generalized, for there seems to be no method of guaranteeing the future of such operations as we now accept (*21*:125). The final reference is always to particulars in their contexts (*20*:56), and definition in terms of unique operations is the only *safe* procedure in physical situations (*27*:255).

3. Need for Generalizing Operations

It is difficult to reconcile these opposing tendencies in Bridgman. Apparently operations both differ from one another and resemble one another. As actually performed, operations are unique, and concepts defined in terms of these operations should preserve this uniqueness. But where operations resemble one another closely this uniqueness may be disregarded and two similar operations may be said to define the *same* concept. The simplest way to express this fact would be to say that in all conceptual definition there is implicitly present an operation of *gen-*

eralization. Only by virtue of this operation are we permitted to disregard both the accidents of time and space, which determine the uniqueness of *each particular operation,* and the accidents of the specific characteristics, which determine the differences between *kinds of operations.* Where the operations both resemble and differ from one another we may neglect the differences in favor of the resemblances—an act which is made possible by our ability to generalize. Apart from acts of this sort no concepts of any kind would be possible and our scientific vocabulary would be limited to proper names. This may have been one of the reasons for Bridgman's introduction of mental operations. Certainly a mental operation of this kind is not only possible but absolutely necessary if there is to be any such thing as conceptual thinking.

Stevens has tried to take care of this difficulty by admitting that there is a "procedure for generalizing from operations" (63:33). He does not state that generalization thereby itself becomes an operation which necessarily enters into the meaning of every concept. But he admits that the rules for generalization should be stated. Unfortunately, in this respect, he has no distinctive contribution to make. We combine operations when they satisfy the criteria of a class; and the class-concept is defined by the operations which determine inclusion within the class (63:34). We construct classes by "correlating" our discriminations, and in this process a certain latitude is allowed. The result is that no concept is without its halo of uncertainty. But this says only that in order to determine a class we must have a concept, and to the extent to which the concept is vague the class is inadequately determined. This is hardly the unique contribution of operationism.

Marx (63:124) states the issue clearly but does not show

how it is to be settled. "If concepts of a high degree of generality are essential objectives of scientific theory, how can they be obtained through the use of constructs postulated specifically to refer to a particular set of experimental operations?" His only solution, as we saw, is to propose that a particular construct, E/C, be given strict, operational meaning, and then subsequently generalized to refer to an increasing number of different kinds of experimental situations. This must always be done, of course, "with considerable caution." Tolman states frankly that we merely "assume" that a curve which adequately describes behavior response for certain values of the variables will hold of others as well (63:95), and that every curve is itself a concept whose meaning involves explicit reference to a generalizing operation.

Israel (86:260) puts his finger on the precise objections to this procedure. Let us suppose that two metrical operations are shown to be equivalent by virtue of the fact that when performed they yield the same quantities. This argument "involves detaching the concept of quantitative value from its operational meaning and assigning to it the status of an absolute property, quantity of a kind which transcends the methods by which it is determined. Equivalence is a relative concept which demands a point of reference outside of the equivalent items . . . outside of the operations themselves." "By introducing the non-operational construct of absolute quantity the operationist escapes from the narrow limits of his highly restricted doctrine." If I understand Israel here, what he is saying is that the operational theory is inadequate because in order to classify two operations as equivalent we must employ a non-operational concept as the principle of our classification. This seems to be precisely what the ra-

tionalists have been saying for many years when they
have insisted that particulars can be grouped into classes
only by virtue of universals, which are not themselves
completely definable in terms of particulars. Stevens
might reply by stating that the difficulty can be met by
admitting that this "absolute" quantity is itself an operational concept derived by generalization. But Israel's
point would remain, that if this is so the operation of generalization is unique among operations, and its employment is in fact precisely the kind of procedure which the
original operationists, in their reaction against the absolutes of Newtonian physics, were trying to avoid.

The difficulty can be further clarified by analyzing Chapin's procedure in his attempt to give an operational definition of "socio-economic status." It will be recalled that
the concept is first defined verbally in terms of the position
which a family under consideration occupies, as indicated
by cultural possessions, income, material possessions and
participation in group activities. It is then redefined operationally by showing that the values obtained by the
measurement of these "aspects" of socio-economic status
and of certain others, such as education of the mother and
father, exhibit a high degree of correlation. "Socio-economic status" is then *defined* as the correlation of these
values.

But why are just those measured values examined for
correlation? They do not give "very nearly or perhaps
identical results," but only results which have a fairly
high degree of correlation. How high must the correlation be to warrant grouping the values together as measurements of the same thing? In the case of tapeline and
triangulation measurements of the length of a field there
is an obvious truth in saying that they are measurements

of the same thing; but this thing—length—must, in Israel's phrase, "transcend the methods by which it is determined." "Length" must *mean* something "over and above" that which is detectable by tapeline and triangulation, and the presence of this "plus-element" permits an indefinite number of later measurements to be included in the definition of the concept.

Consider the simplest case, in which I measure the length of a field by tapeline operations and come out with a certain result. The outcome is a specific number which cannot have the status of a concept, since it is defined only by particular operations performed here and now. The length of the field *is*, then, this number. But now suppose I wish to check my measurement by performing the operation a number of times. Regardless of whether the results are in substantial agreement I can *redefine* "the length of this field" as the aggregate of the numbers obtained, or as the mode, or the median, or the average, or some other function of them. This is still a particular, since the aggregate of numbers is not a class but a collection. Now I take the next step. Once more I *redefine* "the length of the field." However, I include not only the operations actually performed but also certain others, *not at the moment specified*. Why are they not specified? Simply because, not being omniscient, I cannot foresee what they may be. This means that the expression "the length of this field" has an element of vagueness—something "absolute," in Israel's sense, or at least something which cannot at the moment be operationally defined. Yet I prefer to leave the expression in this vague form in order that I shall not be required to redefine it every time I discover a new mode of measurement.

True generalization could be achieved only in case I

were to proceed to *any* field, and thence to *any* area. This would involve a further element of vagueness, since, as Bridgman himself points out, measurements of the very large and the very small frequently require new types of operation. But *unless* and *until* these operations are specified the concept is vague, and *if* they are specified the concept becomes restricted in range, and any later operation, even if it is found to produce similar results, will define a different concept.

An excellent example of the *reductio ad absurdum* of this basic insistence on particulars can be seen in a proposal by Franz Adler (*1*:438-44). He suggests that we attempt an operational definition of a certain concept which, in order to avoid confusion with any other concepts, we shall call "C_N." This score is the sum of the values obtained when an individual fills in a questionnaire which asks how many hours of sleep he had on the previous night, how long he estimates his nose to be, whether or not he likes fried liver (plus 1 for yes, minus 1 for no), how many feet he judges there to be in a yard, and so on. The refined C_N rate is then to be computed from this by taking the test daily for "as long as one can take it" and employing an elaborate formula involving C_N, the number of weeks the test is repeated, a correction constant, and so on. The resulting function is, as may be expected, a very complex one containing radicals, exponents and summation signs. Thus it has all the earmarks of a carefully derived scientific concept.

Now what may we do with this concept? It is, says Adler, pure nonsense. Yet "the test measures C_N and C_N is what the test measures. We are confronted here by a seemingly closed system . . . still C_N does not make sense; we are unable to form a concept of it" (*1*:439). This is typical of the procedure of the operationist. "Where the operationist writes about a new phenomenon he devises a measurement, then defines the phenomenon as what is measured by his measurement. Thus he gains an advantage over anybody who might question either his measurement or his concept. This provides an absolute safeguard against criticism" (*1:*441). But what, then, of the claim of the operationists that operational definition provides a tool for differentiating between concepts which correspond to reality and those which are merely verbal? The concept of *age* as applied to a man certainly has both operational definition and reference to reality; so also does the concept of *height*. Now suppose, following the suggestion of Hempel (*49:*46), I multiply a man's height by his age, getting a number which I then call his *hage*. This is operationally defined, but does it have any reference to reality? Possibly. But now let me take the square root of the logarithm of this number. This is operationally defined. Does it have any reference to reality? One would find it hard to say. Even if I were to discover that every man, measured in this way, possessed a unique number, I should be tempted to look upon it more as a device for locating him, like his social security number, than as an indication of any "property" which he possessed. In fact, it looks suspiciously like some of the verbal concepts which the operationists hoped to eliminate by their new method.

One suggestion which Bridgman makes indicates that he may have seen the problem which is here involved. In discussing the meaning of the word "sum," he points out that it is commonly taken to mean the arithmetical sum of *numbers,* but that it tends to be carried over into the area of kinematics and applied to the sum of *velocities.* How-

ever, here the meaning is significantly different, and we
are confronted with one of two alternatives: we can restrict the meaning of "sum" to its original connotation,
and invent a new word, say "addit," for the sum of velocities; or we can retain the word "sum," using it henceforth
to include *both* kinds of operation but realizing that it has
now become ambiguous. This extension of a word from
an originally simple situation to a more complex one in
which only *some* of the original connotation is retained, is
a characteristic trait of language. He concludes by stating that common usage prefers ambiguity and a small
number of words to clarity and a greater number (*20:*19).

4. Conflict Between Clarity and Generality

Bridgman has here placed his finger on the real problem, but he has not analyzed it correctly. In our use of
concepts we *are* presented with something like a conflict
of interests. On the one hand, we want our concepts to
be *precise*. This requires restricting their range of application by specifying all of the properties which must be
possessed by their instances, or, in the language of logic,
decreasing their denotation by increasing their connotation. The limit to this process is the replacement of the
concept by the proper name, which has the maximum of
exactness because it refers to a single instance. The pen-
alty which we pay for this precision is the introduction
into our language of a distinct name for every individual
object. But, on the other hand, we want our concepts to
be *general;* we wish to be able to talk about *kinds* of things
without being required at the same time to specify the individual instances. This demands that we disregard specific qualities (decrease connotation) in order to give
wider scope to our symbols (increase denotation), and

permits us to reduce the number of words in our vocabulary. But it does not commit us to ambiguous symbols, as Bridgman maintains. General symbols are not necessarily ambiguous; the word "color" is no more ambiguous than is the word "red." There is, in the word "color," a loss of specificity, and this makes it appear vague, though *how* vague it is depends on how carefully the generalizing operation is performed. Since it is general, its use may be inadvisable in certain situations where we wish to be specific. But to dispense with the word altogether would be absurd since there are many situations in which we do not have a specific reference in mind, and the general term here serves our purposes admirably. Bridgman's empiricism takes the form of a preference for precision over generality. He believes general symbols to be vague, ambiguous, and definable only by "verbal" operations which involve merely an unending substitution of one word for another. Since particulars alone are real, a definitional route which goes only from words to words cannot produce clarity. He seems to have disregarded entirely the fact that generalization starts with particulars, as does any physical operation, and is therefore perfectly consistent with the empirical outlook.

The point may be made clear by considering proper names rather than concepts. Let us give the name "Socrates" to all the events which happened to this man up to and including his trial at Athens; then the name will mean simply these listed events—his birth, his childhood, his marriage to Xantippe, and so on. But if we do this we clearly cannot say that it was Socrates who drank the hemlock, for this was not listed among the events to which we attached the name. We shall then have to invent a new name for the man who drank hemlock and died. We could avoid this difficulty by defining "Socrates" in terms of the events listed, *together with* certain other events in the future connected with these by relations of resemblance and causation. But then the name would become vague.

Now let us pass to concepts. The word "red" would commonly be said to be defined in terms of this red, that red, and the other red which I have experienced and which I now recall. If this is the case I cannot ever experience any new reds, for these could not be characterized by the concept no matter how closely they should resemble the former cases. The only way to permit reference to future reds is to define "red" as a definite collection of past reds together with some other events, still in the future, which bear to the past events a certain degree of resemblance (usually not specified). If we adopt the former method our concept is precise, but we must invent a new concept to apply to the newly discovered events. If we adopt the latter our concept may be applied to the new cases, since it is vague in its reference; but we thereby avoid the need for inventing a new word.

Finally, if we pass now to operationally defined concepts, we can see that the problem is essentially the same. "Intelligence" may be defined either in terms of existing intelligence tests, in which case the word is quite precise, or it may be defined in terms of something vaguely recognized as "intelligence"—something revealed by the existing tests but not to be restricted to them since other tests for the "same" thing may be found in the future. Suppose, in fact, that someone proposes measuring intelligence by the use of brain waves. If we have previously defined "intelligence" in terms of existing tests, we clearly cannot call the new device an intelligence test.

We must therefore invent a new word. But if we have previously defined "intelligence" more vaguely, we may apply the word to the new tests provided we feel that the results of the brain wave method and those of the question and answer method exhibit a sufficiently high degree of correlation. By choosing the former method we are able to speak more precisely, but we must multiply terms indefinitely, the extreme of which would be a language consisting wholly of proper names. By choosing the latter we open ourselves to the charge of vagueness, but we both keep our vocabulary within manageable limits and provide ourselves with *general* terms. A narrowly conceived operationism seems definitely committed to the former.

5. Disregard of "Things"

There is a second criticism which may be directed against Bridgman, and against at least some of those who have followed his lead. While this is closely related to the criticism which has just been discussed it is not, I think, identical with it. Bridgman and many of his followers have tended to disregard, or at least inadequately emphasize, the fact that an operation must always be performed *on* something. This has seldom taken the form of a flat denial that operations need things on which they may operate, or of an assertion that they occur in a complete void. Yet Bridgman's statement that concepts are *synonymous* with operations hardly yields to any other interpretation. Frequently the concession is simply, as in the case of Dodd, to "materials used." But this seems hardly adequate. Even the doctor does not simply perform operations; he operates on somebody and on some portion of his anatomy. What is more important, the particular operation performed seems to be determined partly, if not wholly, by the nature of that which is operated upon. Any object is the potentiality of all acts that can be performed on it, and the introduction of these operations transforms the object as potentiality into the object as actuality. A piece of sodium, for example, is not only a silver-white substance, but the potentiality of sodium chloride, sodium hydroxide, and all other compounds into which it may enter by proper chemical action. In the most general sense, an operation is any act performed on something to produce something else or at least a change in the original thing operated upon. And what is produced is clearly the result both of what was operated upon and the operation performed. We perform certain operations in measuring space and certain others in measuring time. This difference is based on the fact that we are immediately presented with the "spread-out" character of space, on the one hand, and the "passage" of time, on the other, and we thus see that the use of micrometer screws, tapelines and light signals is suitable for measuring the former, and hour glasses, clocks and the movements of the heavenly bodies are appropriate devices for measuring the latter. This seems to have been recognized by Bridgman when he says that "the physical essence of time is buried in that long physical experience that taught us what operations are adapted to describing and correlating nature" (14:79). To delve into this, he admits, would go beyond the scope of operational definitions and, perhaps, "beyond the verge of meaning itself." Thus space and time mean something independently of operations, and it is only because we grasp these meanings, however vaguely, that we are able to invent methods for rendering them precise through operations. But if there is meaning beyond the scope of operational definition one may well ask whether the operational theory can then be adequate.

To this question Bridgman has, presumably, an answer. For example, in discussing the claim that the relativity theory reduces simply to "pointer readings" or "coincidences," he points out that if the equations are to have content there must be some sort of physical happenings whose qualities are not determined wholly by the mathematical framework. Neglect of these happenings produces the absurd statement, often made, that there is nothing in nature except pointer readings, or that *Alles ist Coincidenz.* "Let anyone," Bridgman says, "who maintains that there is nothing in nature except pointer readings or coincidences, engage to reproduce the situation that gave rise to the pointer readings in terms only of the framework and the pointer readings themselves" (*19:75*). Or, again, in discussing whether "intelligence" can be defined as "what intelligence tests measure" he points out that any "definition of phenomenon by the operations which produced it, taken naked and without further qualification, has an entirely specious precision, because it is a description of a single isolated event, without even the existence of a criterion to determine when it recurs or whether the description of the event is complete" (*86:248*). Initially, he says, we use the word "intelligence" to describe certain aspects of behavior, of ourselves and of others. Since the word is much too vague to be precisely applied we set about to find some simple procedure which will make more exact what we vaguely feel to be involved in the concept. But how do we set about to do this? We must be sure that intelligence "recurs"; this means that the concept cannot be so vague as to prevent our recognition of repeated instances. But mere recognition of repeated instances does not suffice. For "operational definitions, in spite of their precision, are in application without significance unless the situations to which they are applied are sufficiently developed so that at least *two methods* are known of getting to the terminus" (*Ibid.*, italics mine).

Here Bridgman seems to be saying quite clearly that intelligence is a *thing* which we at first vaguely recognize in ourselves and in others, and which we then try to render precise by noting, first, its recurrence, and, second, its testability by at least two methods having some kind of correlation. It is perhaps not important for Bridgman's general position that the two methods for measuring intelligence may be quite different from one another, while the two methods for measuring the length of a field may closely resemble one another. In either case the two methods must "give very nearly or perhaps identical results." But having measured intelligence in this way we have an operational definition of the term which we then substitute for the old definition on the grounds that it is more precise and easier to apply. This is the *concept*, intelligence, to whose definition we were led by the vaguely recognized *thing*, intelligence.

But many of Bridgman's statements seem to contradict this. For example, he points out that the definition of "intelligence" in terms of what the tests measure begs the question, for by using the word "what" it assumes "the repeated application of the test and the discovery that the results of the test have the properties of a 'what'" (86:248-9). And the word "what," Bridgman says, is simply an accident of language and should not commit us to a Platonic realism (5:240). In these statements Bridgman seems to be denying that there is any "what" which originates the problem in the first place. Thus

Bridgman seems both to affirm and to deny the need for "things" on which operations may be performed.

6. Presumed Identity of Nature and Knowledge of Nature

If we generalize this difficulty we can see that the denial of "things," or, at best, the uncertainty as to their status, is responsible for some of the absurdities and confusions in the operational point of view. If there are no "things" but only operations, then presumably there is no nature but only our knowledge of nature. Bridgman and Stevens seem willing to accept this strange consequence. Says Bridgman, "From the operational point of view it is meaningless to separate 'nature' from 'knowledge of nature'" (*14:62*). Says Stevens, "A statement about a datum is a statement about a construct . . . we had best dispense with the distinction between data and constructs . . . there are only constructs—of various orders of complexity and inference" (*84:100*). The similarity in these two beliefs can be readily seen if we add that for Stevens the immediately given experience is defined operationally in terms of differential responses of the reacting organism. In "placing discrimination at the basis of all science we allow it to usurp the position formerly enjoyed by 'experience' or the 'immediately given.' Does this substitution mean that experience and elementary reactions are equivalent? It does, precisely" (*84:95*). Thus all constructs—which are indistinguishable from data—are behavioristically defined, and there can be no operational difference between "that which" produces the behavior reaction and this reaction itself.

It should be noted that Bridgman says merely that on operational grounds we cannot *separate* nature from our knowledge of nature, but that Stevens says we cannot *distinguish* them. The two men are not therefore asserting precisely the same thing unless Bridgman is willing to admit that inseparability implies indistinguishability. But let us examine Stevens' statement in the form, "Nature is identical with our knowledge of nature." Presumably the proof of such a statement would have to establish that it is: (1) analytic, (2) empirically confirmed, or (3) pragmatically useful.

(1) I am unable to see that the proposition can be analytic, except by the introduction of arbitrary definition. One could, of course, set up a postulate system which would make it analytic. One might say, in fact, that the very postulates of the operational point of view—supposing them to have been formulated much more precisely than has been the case up to the present—do exactly this. Similarly, the non-operationist (who has been even less successful in formulating his postulates than has the operationist) would presumably find the statement contradictory. The point is that as the words "nature" and "knowledge" are ordinarily understood it is neither obviously true nor obviously false that nature is identical with our knowledge of nature, and only the adoption of more or less arbitrary definitions of these two terms would make it either the one or the other.

(2) If the proposition is empirical it may be one which is confirmed by direct inspection of facts, or one which is indirectly confirmed in terms of its consequences. Being a general statement, *i.e.,* a statement to the effect that nature is *always* indistinguishable from our knowledge of nature, it could not be confirmed by direct inspection of any particular fact of cognition. It must therefore be an hypothesis, progressively confirmed in terms of its impli-

cations. If this is true we ought in a class of situations like, say, our knowledge of trees, to be able to decide by some sort of direct inspection whether this knowledge is indistinguishable from the trees themselves. But the answer in this case cannot be decisive, since it again depends on how "knowledge" is to be defined. If the word is taken to include any "awareness" of an object in the broadest possible sense then the statement would be confirmed, since every case in which we "get" nature is also a case of our knowledge of nature, *i.e.*, to "get" nature is to know of its existence. On the other hand, if "knowledge" is opposed to something which can be called "presentation," and consequently restricted to symbolic or conceptual interpretation, then the statement in question would be disconfirmed. We often have "knowledge about" things when we are not "presented" with them, and we even find, in fact, that when we are "presented" with them we can still distinguish our knowledge from that which is presented. This certainly means that what is to be known, and our knowledge of it, are distinguishable.

(3) To establish the pragmatic validity of the statement would not be an easy task. This is due to at least two considerations. In the first place, the operationists are themselves divided as to whether the statement or its contradictory is the more useful to believe as a working basis for science. We have already seen how strong the pragmatic motivation is in the operational attitude. All operationists seem disturbed by the failure on the part of science to advance as rapidly as it should, all are searching for the reason for this, and all seem to find it in the employment of methods which allow vagueness, "metaphysics," non-empirical hypotheses and other non-operational elements to enter as presumed concepts of explanation. Some op-

erationists, however, while insisting on the desirability of precisely defined concepts, are willing to admit that vague ideas, imaginative hypotheses, models and other devices have a certain temporary role to play in science. In the case of Marx, for example, hypotheses and other ideas expressed in the language of common sense are important in the early stages of the development of a science; these should, however, be replaced by operationally defined concepts as soon as possible. So also in the case of Lundberg, whose admission of the notion of "that which" evokes responses might well constitute an acknowledgment that nature is distinguishable from our knowledge of nature; he would presumably say that if "nature" is that which evokes the responses which we express by "knowledge of nature" the two must be distinguishable, at least in theory. To be sure, he would deny that any assertions can be made about this "ultimate nature," since all such hypotheses would be unverifiable and therefore outside the sphere of science. Presumably, therefore, his willingness to admit the existence of "that which" evokes responses, in distinction from the responses themselves, is pragmatically motivated.

In the second place, the pragmatic justification of the statement in question would be difficult by virtue of the fact that there is not just one goal of science but many. Even if we exclude the practical goal, *e.g.*, the production of gadgets, the elimination of disease, the increase in material comforts and the like, the theoretical goal itself is variously conceived. In the very broadest sense, of course, we want knowledge which will be as accurate and comprehensive as possible; this principle would certainly be accepted by operationists and non-operationists alike. But in the achievement of this goal we may employ methods and procedures which are not only diverse but may even be conflicting. For example, *clarity* is obviously an important criterion for scientific ideas. But clarity can be achieved only to the extent to which there is "agreement" between nature and our knowledge of nature, and perfect clarity would presumably require that these be identical. On the other hand, predictability of truths not yet known is another important criterion for scientific ideas. But this would require a sharp differentiation of nature from our knowledge of nature, since it would demand a distinction between *potential* knowledge and *actual* knowledge. As a result, the pragmatic justification of the statement in question seems also to fail. If clarity and predictability are conflicting but legitimate goals of knowledge the devotees of clarity are entitled to insist upon the identity of nature and our knowledge of nature and the devotees of predictability to deny it.

But there is another line of attack against the operationist's claim of the identity of nature and our knowledge of nature. This involves the demonstration that the operationists who explicitly accept this identification implicitly accept also its contradictory, thus exhibiting an inconsistency in their own thinking. We have already seen that this is the case with Bridgman. Take also the case of Stevens. He points out that operations may be of various orders of complexity, extending from elaborate experimental procedures at one end to simple discriminatory responses at the other. Among these the discriminatory responses are the most fundamental (84:94-5). In fact it is precisely these responses which define "experience," for he says that by a sensation is meant a "class of reactions." "What is important here is that from the point of view of science we never find *red* as such, we find only such situations as *man sees red,* i.e., an organism discriminates. Thus the *sensation red* is a term used to denote an 'objective' *process* or event which is public and which is observable by any competent investigator" (83:524). But how about the investigator himself? "As a psychologist he observes the reactions of another human being and what is even more interesting, he may himself be observed by another experimenter who may in turn be observed so that we pass . . . into a potentially infinite regress." But then Stevens says—and this is very significant—"we must disrupt such a regress by stopping it arbitrarily at some point. In other words, in admitting any item to the body of scientific knowledge, we assume at some stage an independent experimenter whose qualifications to observe and record data are not in question" (84:95). But *what* is this experimenter who must be assumed? He is something which is not operationally defined (though he may be operationally definable if discriminated by another experimenter) and which must be assumed or taken for granted at some point in order that science may go on (47:312). This is the fate of all relativisms. If A is defined only by its relation R to B, and if ARB is then defined only by its relation R' to C, and so on; then we are confronted either with an infinite regress, in which case nothing can be defined, or we must arbitrarily stop the process and admit that some particular relational complex, say, (ARB) R'C, is not itself relative to anything else.

This inconsistency in Stevens' position can be clearly indicated by another quotation. "Any attempt to define the term experience operationally or point out what, concretely, is meant by the philosopher's 'given' discloses at once that the discriminatory reaction is the only objective, verifiable thing denoted" (84:95). This says unmistakably that the discriminatory reaction is more objective than that which produces it. But if a psychologist observes the reaction then *his own* discriminatory response becomes more objective than the reaction which he is observing, and so on for other psychologists observing *him*. Now the only way to avoid the infinite regress is to suppose some reaction which has its character determined without reference to any other reaction toward it. But, if we are obliged at some point to suppose something definable without reference to any reactions, why not identify this with the "philosopher's 'given'" and thus avoid creating a regress which we have to terminate arbitrarily? If the arbitrary termination is demanded by the need for supposing *something* as given without reference to a reaction, it is simpler to make this a sense-datum rather than a reaction to a reaction to a sense-datum. If nature is then identified with the given, and our knowledge of nature is identified with any discriminatory or other operational reaction to this given, one can no longer claim that nature is indistinguishable from our knowledge of nature. Nature will be "that which" originates our knowledge, or nature will be the potentiality as over against the actuality of knowledge. In the words of Feigl (who does not adopt the extreme operational point of view on this matter) "things are . . . what they are known *and* know*able* as" (86:257). To admit the concept of the know*able* is inconsistent with a thoroughgoing operationism.

The final objection to this claim that there is no distinction between nature and our knowledge of nature applies mainly to Bridgman's statement that concepts are *synonymous* with operations, though it also applies to all those operationists (Stevens, Lundberg) who tend to relegate to a position of insignificance "that which" evokes the operational response. Since, as we have seen, an operation is always performed on something one would expect a concept to be defined in terms of two entities—the thing operated upon, and the operation performed. But if the thing operated upon is either denied, as in the case of Bridgman's definition, or expressible only as "that which" or as an x (59:15), what could be meant by an operational definition? Consider, for example, how, if this were the case, one would define the very concept "operation" itself. The concept "operation," of course, refers to an operation. But the operation is "that which" is referred to by the word; there is no other operation involved in its meaning, except that of generalization, and this is involved in *all* concepts. The point is that the concept "operation" is *explicitly* operational, since it means an operation, but the concept "length," for example, is only *implicitly* operational, since an operational act is involved in its determination. Or, to put the point another way, the concept "length" and the concept "measuring" are fundamentally different in character and cannot receive the same kind of operational definition. Measuring is an operation but length is not. Concepts which are explicitly operational are those which are represented linguistically by words ending in "-ing," such as "counting," "experimenting," "generalizing," and "inferring," with correlative forms sometimes ending in "-ion." Concepts such as "length," "mass," and "time" do not mean operations explicitly, though they may require the use of operations in their definitions.

If the operationists are to be interpreted literally in what they say on this matter they are implicitly committing themselves to an activistic metaphysics which, while not an impossible world-view, is indeed a strange one for a group who tend to scorn metaphysics of any kind. Yet many of their statements admit of no other interpretation. Stevens' claim that the "given" *is* simply the discriminatory reaction is unequivocal; reaction is a kind of action, and action is activity; hence the world (as that which can be given) is mere activity. Similarly, when Bridgman says that length *is* measuring, science *is* *s*ciencing, and intelligence *is* testing, he is taking the same position. On other occasions he says that our methods of handling the world have a greater stability than the external world itself (27:257), and that one important aspect of the operational point of view is to see the world in terms of activities rather than in terms of things (28:4). This is not, of course, an out-and-out metaphysics, since Bridgman does not say that the world contains only operations. But when he says that an analysis of the world into activities "goes beyond" an analysis into objects (28:5), there is a strong suggestion that an activistic interpretation of the world would be more basic than one which is not.

7. Failure to Provide a Classification of Operations

The third criticism, which applies in general as well to the followers of Bridgman as to Bridgman himself, is the failure on the part of the advocates of this position to provide us with an adequate classification, or even with an adequate list, of the main *kinds* of operations. A possible exception to this is Feigl who, as we have seen, suggests that there are five main types. In the context of this listing, however, he does not make any claim that the list is exhaustive, nor does he discuss in any detail the distinctions between the various kinds (86:257). Dodd admits that there may be qualitative as well as quantitative operations, but neither he nor Feigl tells us what a qualitative operation is. Stevens calls attention to the need for recognizing as the most fundamental operation that of discrimination (83:524), and he agrees with Feigl that mathematical operations should not properly be included in the operations of operationism (63:28). Other psychologists (63:114-22, 71-81) have attempted listings of "types of theoretical constructs," and if these may all be presumed to be operationally definable the listings may be taken to represent kinds of operations required to define the respective constructs. But, as we have seen (Chap. IV), the confusion with regard to kinds of postulational symbols (theories, hypotheses, constructs, intervening variables) makes such listings practically useless. The distinction between methods of inference and methods of construction seems to correspond to what many operationists would characterize as the distinction between the vague, non-operational techniques supposedly employed by the metaphysicians and by early scientists, and the precise, operational techniques which are found in the "approved" sciences of today. But if this is true then certain types of postulational symbols would be non-operationally defined, and a list of kinds of postulational symbols could not be taken as a list of kinds of operations.

As we have already seen, Bridgman admits mental operations in addition to physical ones, and he suggests that the former can be roughly divided into verbal operations and pencil and paper operations. None of these kinds can be sharply *differentiated* from the others, and none can be *separated* from the others (21:123, 129; 27:256). With regard to the possibility of separation he says that they are so intertwined, and so reinforce and supplement one another, that it would possibly be meaningless to attempt

such a thing (29:160). He mentions "directed" operations, *i.e.*, operations involved in "understanding," and
thereby suggests that there may be operations which are
not employed in understanding. But he has given us no
analysis of understanding. For example, we cannot distinguish between good and bad operations (21:126) or
between useful and useless operations (86:248) unless we
have a clearer conception of what the goal of understanding is.

8. Bridgman's List of Operations

Perhaps, therefore, the best way to approach the problem is to list the various kinds of operations which Bridgman mentions. Since he nowhere presents such a list,
and, indeed, at one point suggests that it would not be
profitable to attempt an exhaustive classification of the
non-instrumental or mental operations (25:483), I cannot assume that the enumeration here given is complete.
Doubtless some important kinds have been omitted. First
are physical operations. These are also called "manual"
(21:128), "instrumental," and "laboratory" operations
(29:258). The most commonly cited example is measurement. Non-physical operations are all mental, and these
are subdivided into verbal operations and paper and pencil operations. Mental operations are characterized by
the fact that they "have no necessary physical validity"
(19:11). They can also be called "mental experiments,"
and they exhibit a world in which free invention is possible. Analogy, for example, is a mental operation (29:-
260). Mental operations are subject to the restriction that
they must be capable directly or indirectly of "making
connection with instrumental operations" (*Ibid*). Verbal
operations are mental operations which are illustrated by
asking oneself questions (What shall I do next? Who said

that? Would I say thus and so in such and such a situation?), and by making verbal substitutions (*21:125*). In our use of verbal substitutions we frequently discover to our dismay that there are verbal regressions which are unending. These must be called "unresolvable" since they do not end in something non-verbal. Paper and pencil operations are best illustrated by mathematics, but they include any manipulation of symbols, regardless of whether or not these are the conventional symbols of mathematics. Logic, for example, involves paper and pencil operations; these include the drawing of conclusions by syllogistic reasoning and by induction, and the invention of logical laws such as the law of excluded middle (*19:36; 20:47*).

Such a classification of operations, while suggestive, leaves much to be desired in the way of precision and exactness. Its inadequacy is clearly disclosed when we try to fit into it certain operations which Bridgman apparently admits into his scheme without telling us exactly how. Observation, for example, is an operation. Bridgman says that the subject matter of all attempts at understanding is activity of one sort or another, either in the perception and recognition of sense impressions, or in the performance of deliberate physical manipulations or of deliberate thought (*21:115-6*). Again, he says that testing is an operation; this deliberate act may be simply placing oneself in such a position that his sense organs may be acted on and then paying attention to the resulting sensations (*86:246*). Presumably this is a mental operation, but it is not strictly either verbal or paper and pencil. Furthermore, if perception is an operation then at least some operations always involve a "what," a something which is operated upon; for perception in the absence of anything to be perceived is meaningless. And to admit that in "recognizing our own feelings" (*20:75*) we are performing an operation, is to grant the need for a referent for operations; we could hardly observe our own feelings if they were not there to be recognized.

In addition, explanation is presumably also an operation (*14:37 et seq.*). But if so it would have to be mental, and at least partly verbal. Furthermore, the essence of explanation consists in reducing a situation to elements which are so familiar that we accept them as a matter of course, and our curiosity rests (*Ibid.*). But not all explanations are "good": to explain by "metaphysical" operations (*19:10, 35*), "idealizing" operations (*Ibid.*), "mystical" operations (*20:27, 77*) or by the operation of "being used with certain verbal forms" is in each case bad, and to explain by "models" and "constructs," while useful, is a dangerous method (*14:53*).

An attempt will be made in the concluding chapter of this study to eliminate some of the confusion in this list of operations by presenting an alternative list whose members are, it is hoped, somewhat more precisely defined. Bridgman wants to justify the operational approach by pointing out the obvious fact that in the broadest sense operations are completely inescapable; everything that we *do*, in fact, is an operation. But many acts which we perform have nothing whatsoever to do with our cognitive activities; consequently operations should be restricted to those which are useful in explanation. Now these unquestionably involve something more than merely physical operations, since we must also observe, invent symbols and then manipulate these symbols by verbal operations in order to get them into the form which will permit us to say that they explain something. And this means that explanation as an act is normative or purposive, and can only be formulated in terms of what we consider to be the goal of the scientific endeavor. Bridgman's difficulty, as well as that of many of his associates, lies in the fact that the discussion of operations, and of their presumed justification, is carried on without benefit of a general theory of knowledge which would define the goal of the intellectual process. In Bridgman's case this is clearly indicated in his continual shift between the pragmatic theory of truth and the correspondence theory; it should be obvious that operations which are quite legitimate according to one theory are not according to the other. What should be even more obvious is that the conflict between the operationists and the non-operationists is itself due to fundamentally opposed theories of knowledge.

9. Failure to Distinguish Symbolic and Non-Symbolic Operations

The final criticism is the failure of most operationists clearly to distinguish operations which are symbolic from those which are not. Strictly speaking only symbolic operations are cognitively significant. By "symbolic operations" are meant any operations which "produce" symbols or which give meaning to symbols. For example, a denotative gesture of pointing to an individual and saying, "That's Smith" is a symbolic operation employed to give meaning to the name "Smith." Similarly, an ostensive definition of "red" may be given by pointing to a group of red things, and adding that what is meant by "red" is these things together with others bearing a certain kind of resemblance to them. Again, an operational definition of the temperature of a gas may be provided by inserting a thermometer in it and then observing the number coinciding with the height of the column of mercury; or an operational definition of the slope of a line by determining the tangent of the angle which the line forms with the horizontal. In some of these cases the operations are dominantly physical and in some cases they are not. But it is in the physical cases that there is the greatest danger of confusing symbolic operations with those which are non-symbolic. Take the case of measuring the temperature of a gas. This can be looked upon as a purely non-symbolic operation, having no cognitive significance. It involves setting up a complex causal situation consisting of the gas and the inserted thermometer, and then noting what effect is produced. Now what is important is that the effect is itself simply another physical situation—the position of the mercury in the tube. Even the number on the scale is merely a physical event. Taking the temperature of a gas is exactly like pushing an electric switch to turn on the light, or driving in a nail with a hammer, or baking a cake. No symbolic operations are involved, and the acts might be performed more or less unconsciously. In all such cases we are neither trying to *learn* anything nor attempting to *create symbols* which may later have cognitive significance; we are simply trying to produce certain desired results by introducing the proper causes.

What, then, gives such an operation cognitive significance? The answer is simple and clear-cut. The event which is produced by the operation must *refer* to that which was involved in its creation in that unique way which is characteristic of all *symbols*. Symbols are a special kind of *sign*. A sign is defined as that which has the property of referring to, or indicating, something else; this "meaning relationship" is probably unique and indefinable. But it can be readily exemplified: an overcast sky means rain, a flag at half-mast means the death of an important figure, the reading on a thermometer means that something is hot or cold, the fact of the light being on means that someone has pushed the switch, and the existence of a cake means that someone has baked it. Whenever any two physical events are frequently associated either may become the sign of the other, and the presence of the one may lead a person to think of the other. But not all *signs* are *symbols*. A sign becomes a symbol when it is taken away from the physical situation and made into an instrument of communication; this involves making it a part of a language system having syntactical, semantical and pragmatical rules. In the great majority of cases it also involves replacing the physical event by an arbitrarily chosen symbol, such as a word or a number. In pictorial languages the character of the physical event is retained in the symbol; for example, one could easily construct a language in which heat would be represented by the picture of a thermometer with the mercury standing at a high point on the scale. But in most modern languages this element of resemblance between the symbol and its referent has been lost, and the associative tie is established simply by learning the language.

10. Three Kinds of Operations

This means that there are really three kinds of operations. First, there are physical operations which produce physical events; these are of no direct cognitive significance since neither the producing events nor the produced events have the capacity to refer to one another. Second, there are physical operations which produce events having the capacity to refer back to those events involved in their creation; such events may be said to be the "signs" of the causal events involved in their creation. Finally, there are operations which may involve physical aspects but are dominantly non-physical in character and which eventuate in linguistic symbols; these symbols have the capacity to refer to the events which were involved in their creation, but they have been freed from their physical basis and have become parts of larger symbolic schemes having internal relations and rules for usage.

The distinction between the three kinds can be illustrated by the three interpretations which may be given to the claim made by many physicists that science is concerned only with "pointer readings." In the first sense of the word "operation," pointer readings are simply physical events connected with other physical events by the principles which are exemplified in the construction of the recording device; in this sense clocks do not measure the flow of time but are simply rhythmic objects which differ from the movement of the sun only in the fact that they have been created by man. In the second sense of the word "operation," pointer readings are physical events which may signify other events; for example, the change in the position of the hands of a clock may signify the passage of an hour. Finally, pointer readings may be abstracted from the physical objects with which they are associated and given a symbolic character which permits them to be used as devices for communication; the word "hour," for example, in contrast to the physical change in the position of the hands, is such a symbol. In the case of the thermometer the distinction would be between a mark, such as 85, which represents an actual position on a scale, and the symbol "85 degrees Fahrenheit" which is part of an extensive linguistic system.

Now if one accepts the claim, which seems to be agreed upon by all operationists, that operations provide a technique for defining concepts, only the third type of operation will be relevant to the problem; for physical events are not concepts even when they have the capacity to act as signs of other events. Concepts are linguistic elements having a unique referential capacity; they are, for example, *general,* which no physical event is. The difference can be illustrated by distinguishing between the red sky in the evening which may *signify* clear weather on the ensuing day, and the words "clear weather tomorrow" which may *symbolize* the same fact. Only symbolic operations, together with those physical operations which may be associated with them, are of significance in the operational point of view in so far as this is considered to be a theory of knowledge or of method. The function of operations is to define, and the only thing we want to define is a symbol, be it a proper name, a concept, a measured value, an hypothesis or any other symbolic element.

11. Operations as "Producing" Something

The point which is here involved can be approached from another angle. Let us take some of the common formulations of the operational point of view: Concepts are to "be defined in terms of the operations that *produce* them" (Marx 63:20); an operational definition specifies the procedure which must be carried out not only in order to "identify" a case of the definiendum but also in order to "*generate*" it (Dodd 39:483); "a *recipe* for a chocolate cake is an operational definition of such a cake" (Lundberg 60:89). In all these quotations (the italics are mine), there is the notion of operations as producing or creating something. In the case of Marx that which is produced is clearly stated to be a concept. But in the case of Lundberg it is not a concept but a thing, a cake. And in the case of Dodd, while the word "definiendum" should properly refer to a concept (since the only thing we define, in the strict sense, is a term), nevertheless his pairing of the word "generating" with the word "identifying" at least suggests that he also is talking about a thing, since what we are trying to do is to identify a presumed *instance* of a term, not the term itself.

Now the question is not whether there are significant similarities between producing a cake and producing a concept; in both cases we have given materials, we "operate" upon these materials, and we create a product which is in a very important sense the result of the materials and the operations. The real question is whether emphasizing these similarities at the expense of important differences may not be a source of confusion. Take the case of length. When we measure the length of an object and come out with "27 inches," are we doing something more like baking a cake, or more like defining a concept? The answer depends on whether the "27 inches" is a mark on a scale (a thing) or a symbolic description (a concept). Merely to produce a mark on a scale, or merely to bake a cake, are not cognitive operations (with no aspersions intended on the intelligence of people who measure or people who bake), and therefore should not be called either "definitional" or "operational." But to produce a concept and give it meaning on the basis of materials and operations used is at once cognitive *and* definitional *and* operational.

A possible source of further confusion here is the fact that the traditional logic recognizes a kind of definition which is called "genetic." Adler calls this (1:444) "definition by recipe," and a recipe for a cake would then be a genetic definition of a cake. But one should note that this is very carelessly expressed. In one sense, of course, the operational definition of *every* concept is genetic, for we attempt to show how the meaning of the concept originates in things and operations. But on this interpretation there would be nothing unusual in the genetic definition of the word "cake," since one could achieve this merely by generalizing from cakes. In another sense, however, one could define the word "cake" by beginning with a recipe and materials, then baking a cake, and finally defining the word by generalizing upon the outcome of this activity. This is what is ordinarily understood by a genetic definition, and it involves, one should note, both physical and cognitive "operations." In exactly the same way one could construct a genetic definition of "length of 27 inches" by starting with an unmeasured length, producing a mark on a scale through the action of measuring, and then generalizing the result and making it into a symbol. No doubt one could say that the physical operations in both these cases take on cognitive status by virtue of being included in the total definitional process. But *by themselves* they are neither cognitive nor definitional, and to call them "operational" apart from the supplementary acts by which symbols are created is misleading in the extreme. In this sense, to speak of a recipe for a chocolate cake as an operational definition of such a cake is a source of great confusion.

Chapter VI

GENERALIZED OPERATIONISM

1. Need for General Theory of Knowledge

THE FOREGOING exposition and criticism of the operational point of view has, it is to be hoped, brought to light one very important point. No operational theory may be either justified or rejected without explicit reference to a broader and more general theory of knowledge. In the case of Bridgman this has been clear. He wishes to be an empiricist, and he wishes to be a pragmatist. In support of the former he must insist on precision in concepts and certainty in judgments, but in support of the latter he must demand that the real measure of the adequacy of any concept be its general contribution to the advancement of science. This accounts for his apparent shift from a narrowly conceived physical and metrical operationism in his early writings to a later and more broadly conceived point of view in which operations are allowed to contribute only partly to the meaning of a concept, or (what amounts to the same thing) in which the very notion of operation itself is generalized so as to include a wide array of mental, verbal, and paper and pencil operations. But he seems never to have reconciled completely these opposing points of view in terms of a general theory of knowledge which would attempt to evaluate the respective claims of exactness and utility.

Many of the other operationists have had the same diffi-

(103)

culty. They seem to have been unwilling to face the more general problem, preferring rather to accept an implicit solution, and to allow this to determine their attitude toward operationism. Lundberg departs from strict operationism by admitting something which originates the operations, but which cannot itself be operationally defined; and in insisting on the importance of "traditional meanings" and "intuitive methods," he seems to be giving a place to non-operational techniques. This appears to indicate that he is both favorable to operationism and critical of it, and we cannot decide in the absence of a general solution to the cognitive problem how he would reconcile these opposing claims. Even Chapin, in giving objective status to complicated mathematical functions of more directly derived quantities, seems to be abandoning a rigid operationism; at least an operationism which admits mathematical operations cannot be purely physicalistic. The same is true of some of the operational psychologists, such as Tolman, MacCorquodale and Meehl, and Marx. Tolman, in a more recent writing, has apparently recognized the difficulty in his earlier position. "I am now convinced that 'intervening variables' to which we attempt to give merely operational meaning by tying them through empirically grounded functions either to the stimulus variables, on the one hand, or to the response variables, on the other, really can give us no help unless we can also embed them in a model from whose attributed properties we can deduce new relationships to be looked for. That is, to use Meehl and MacCorquodale's distinction, I would now abandon what they call pure 'intervening variables' for what they call 'hypothetical constructs,' and insist that hypothetical constructs be parts of a more general hypothesized model or substrate" (88:48-9).

This seems to be a frank admission that operationally defined concepts do not have the predictive qualities which are needed for the advancement of science, and that we must therefore replace them by hypothetical constructs which, even if not operationally definable, do have these desired qualities. Porterfield (70:251-3)criticizes the operationists for giving insufficient place to "insight" in science. Alpert (4:857) argues that clarity cannot be the sole standard of science, but that "organizing ability, utility and meaningfulness" are also important. Marx admits that there is a difficulty here when he allows hypothetical constructs in an early stage of science "on a grossly non-operational level of discourse," or even in a later stage insofar as they serve a useful function in research. But at the same time he cannot resist the operational drive which compels him to say that their use cannot be scientifically defended and that they must be transformed into intervening variables as quickly as possible. This seems, again, to indicate a conflict, not clearly formulated, between the empirical and the pragmatic urges in science, with the resolution to be found, as in the case of most operationists, in a rigid adherence to empiricism even though the method, as thus interpreted, no longer seems to be the one which is actually used by scientists.

Many of the same difficulties confront the critics of operationism. While they are agreed that the operational point of view is not tenable as a principle of method, they appear unable to justify this rejection in terms of a more fundamental theory of knowledge. Israel (86:260-1) seems to come closest to recognizing the true limitations of the operational point of view; at least there is suggested in his criticism a more general view concerning the nature and goal of knowledge. That this general theory is not more explicit is, of course, to be regretted. But what Israel appears to see clearly is that many of the operationists are arguing for two conceptions of operation, which are really in conflict with one another. "According to one, a set of operations is a rigidly invariable series of unique operational items, but according to the other, it is a purposive activity directed toward obtaining a certain result with the allowance of variation in the component operations" (86:261). These correspond to what I have called in Bridgman the "narrow," or "empirical," conception, and the "broader," or "pragmatic," one. The conflict may be stated in an equivalent form by saying that concepts are designed to play alternative roles in knowledge. They must refer to that which is already known, hence they must be as precise as possible; but they must enable us to predict a future which we do not yet know, hence they must contain an element of vagueness and uncertainty.

Herbert Blumer's criticism of operationism (12:707-19), like the attacks of Adler (1:438-44), Pratt (71:71-122) and Roback (74:chap. XXV), raises important issues, but constitutes a kind of guerrilla warfare against the position rather than a concerted and well-planned attack. For example, Blumer offers five "solutions" to the problem of how to avoid vagueness in social science concepts. The first is to turn away from concepts to facts; but this misconstrues the task of science. The second is to continue to use the old concepts, and the third is to replace them by new ones; these do not really eliminate vagueness at all. The fourth, operational definition, requires the reduction of concepts to quantitative values and may result in a loss of the "most vital part of the original reference" (12:711). The fifth, arriving at precise definition through critical analysis, is purely lexicographical and has value only as such. These are hardly positive solutions to the problem of vagueness, for they show only that there is no way of avoiding the difficulty. And they do not show, in terms of a general theory of knowledge, either whether precision is to be preferred to vagueness, or, if so, just why this is the case.

Because of the limitations of these expositions and criticisms of operationism an attempt will be made in this chapter to place the operational point of view in the broader context of a general theory of knowledge. Operations are clearly functional in character; they are acts performed for the sake of achieving a certain goal. This goal is knowledge, or, if one wishes to take a more restricted point of view, the kind of knowledge which is sought by science. Now it is obvious that *means* cannot be determined apart from *ends;* hence without a clear-cut conception of what knowledge in its ideal form is, neither the conflict between the operationists and the non-operationists nor the disagreements within the operational point of view as to which operations are allowable and which are not can be resolved. The problem is simply one of determining whether we need operations at all in the creation and development of knowledge, and, supposing that we do, of deciding what operations will be most effective in achieving our end.

2. Demand for Clarity and Certainty

Let us begin with empiricism, stating its main contention in the broadest terms possible. Empiricism maintains that knowing is analytic of experience. This is simply a rephrasing of the more commonly stated thesis that all knowledge is obtained from experience. In calling it an "analytic" act I have tried to place emphasis on the fact that for the empiricist everything that we need to use in the attempt to know experience is itself given in experience; there are no innate ideas, or ideas presupposed by experience but not given in it, or concepts given through a special faculty of reason. If this appears to be too broad a statement, since it seems to render the knower completely passive in the cognitive act, one may point out that this is not the case since the knower must contribute at least three things—the *analytic act itself*, which is his and not nature's, the *symbols* in terms of which experience is to be known, and the *awareness* which is required to differentiate knowledge of experience from experience itself. But the admission of these factors need not destroy the essential empiricism of the position since the knower: (*a*) performs an analytic rather than a synthetic act and thus really "adds" nothing to nature; (*b*) creates not the symbols (their meanings are determined wholly by experience), but only the sign-takens, *viz.*, the written and spoken forms, and (*c*) conceives of his awareness not as creating experience or even as modifying it but only as rendering it knowable in the only sense in which this is possible, since to know anything means, at the very minimum, to be aware of it in some sense.

Now if the adequacy of knowledge were measured simply in terms of *clarity* of concepts and *certainty* of judgments, one could hardly improve upon the empirical theory. If what we know were obtained by analysis from something which were given in immediate awareness, our concepts could be as clear and our judgments as certain as we chose to make them; while there could be dispute as to whether *perfect* clarity could be achieved in the perceptual situation (because of the inferential element in all perception) or whether observational statements could ever be *completely* incorrigible, nevertheless this type of situation would provide the maximum values obtainable in these two dimensions of knowledge.

3. Demand for Range

But others have argued that there is a third dimension of knowledge—extent, or range. We wish to know not merely a small portion of experience with a high degree of clarity and certainty, but a large portion of experience with such clarity and certainty as the methods of knowing permit. Knowledge is not confined to objects given in immediate awareness, where clarity and certainty are at a maximum, but extends to past and future objects, objects located elsewhere, objects "inside" or "behind" those directly given, general connections between objects, and so on. Knowledge of this kind is obtainable only by inference, or by "construction," or by "insight," or by some process other than an immediate examination of the given.

The rationalists have insisted that this process is inference. Inference is possible only because our knowledge is in some sense integrated or systematized; by inference we pass, by means of interconnections within knowledge, from what we know about experience to what we do not yet know. In this way inference provides a method for speeding the growth of knowledge—a logic for anticipating what nature will disclose in other areas and times—and makes possible a more positive direction of the cognitive activity than is the case with a strictly empirical approach.

This is not the place, of course, to review the history of the controversy between rationalism and empiricism. Suffice it to say that many empiricists have reacted favorably to this criticism by the rationalists, and have admitted that within the framework of a strict empiricism knowledge is doomed to progress at an intolerably slow rate. If one is to be a "good" empiricist he will be as completely passive as possible, "listening to nature" and waiting for her to uncover her secrets. Such a method will permit no controlled observation, no experimentation, and no anticipations of what might later be discovered. While there may be something to be said for this approach it is certainly not the method of science today, nor is it a method which is likely to produce rapid advancement in knowledge. As a consequence, many empiricists have sought an alternative method, a method of exploring and prodding nature and of designing experiments and formulating hypotheses which permits us to ask nature certain questions and to expect clear-cut answers. This is precisely the logic of anticipation which the rationalists have aimed to provide.

But now a further difficulty arises. Is this logic of discovery to be *deductive* or *inductive?* If it is the former it will provide a system of rules and principles by which we may: (1) define symbols for objects not given in immediate experience with the same clarity as those for objects which are directly given, and (2) justify propositions about happenings which have not occurred within our immediate experience with the same certainty as those about happenings which have occurred. In other words, if our logic is deductive there need be no loss in either clarity or certainty; in adding the dimension of extent to our knowledge we shall not have to sacrifice these other goals. In the case of an inductive logic this cannot be true.

Since a deductive logic is clearly what we want, let us examine its consequences for our general theory of the universe. If there is to be such a logic of discovery nature as known will have to imply nature as unknown, *i.e.*, the logic will permit us to deduce from any bit of knowledge which we happen to have at the moment propositions which are necessarily true of other areas of our experience. To claim this is to assert something broader than the general thesis of empiricism, which argues that nature contains our knowledge of nature; perhaps nature, as a whole and as potentially known, *does* contain the part which is actually known. The thesis being suggested here is much less plausible, for it argues that the portion of nature which we actually know contains (in the logical sense) the portion which is still to be known, and that we can extend our knowledge merely by logical analysis. But for this to be possible the world would have to be a much more compact logical system than it has ever been judged to be. For the empiricist experience itself would have to have this logical structure, and for the rationalist there would have to be a super-empirical realm of innate or self-evident truths which, if known, would permit the new truths of experience to be deduced.

Neither of these consequences seems compatible with the world as we find it. Experience does not have this logical structure, and the existence of a super-empirical realm known through synthetic *a priori* propositions has at least not been conclusively demonstrated. Our only alternative, therefore, is to suppose that the logic of anticipation is not deductive but inductive. And this means that it can give us neither a high degree of certainty nor a high degree of clarity; it must be venturesome, involving risks and guesses, and requiring us to supplement the satisfying knowledge provided by things which are given to us through our senses by vague and uncertain theories. Such a method ought not, perhaps, to be called a "logic" at all, since it provides no guarantees; it tells us not what the future *must* be but only what it *may* be on the basis of what we already know; and it makes symbols for unobserved objects not perfectly precise but only somewhat less vague than they would have been without such a method. It requires a theory of the universe not as a compact logical system but only as a loose-knit aggregate in which certain parts give "hints" of other parts; hence what we know at any given time will never enable us to deduce what we do not yet know, but will only permit us to make guesses.

Our dilemma now becomes apparent. If we prefer certainty and clarity, throwing out all conjecture and all vague ideas, our knowledge will necessarily remain highly restricted in extent, and we shall have available no technique by which it may be extended. On the other hand, if we prefer knowledge of the widest possible scope, we shall be provided with a method by which this may be achieved, but we shall be forced to admit into the area of knowledge many ideas which are vague and many propositions which are conjectural. We can have certainty and clarity without predictability, or we can have predictability without certainty and clarity; but we cannot have both.

4. Demand for Utility

But now pragmatism comes into the picture. It argues that not only must we determine the methods of knowing by what we want the end-product, knowledge, to be, but we must also determine the very characteristics of knowledge itself by what we want it to do. Knowledge is designed to solve problems, and there will be as many kinds of knowledge as there are problems to be solved and problem-situations to be met. This means that we cannot establish once for all the characteristics which knowledge must possess; knowledge will be of one kind in one situation, and of another kind in another situation. The way in which the conception of knowledge may be broadened by pragmatic factors is illustrated by the following considerations.

Consider the conflict just discussed—that between certainty and clarity, on the one hand, and predictability, on the other. Both the advocates and the critics of operationism seem to have been aware of this conflict, though they have not stated it clearly nor have they recognized its foundation in pragmatic considerations. The operational method was designed to meet the demands for certainty and clarity; propositions must be not only confirmable but actually confirmed operationally, and concepts must be not only definable but actually defined operationally. And if these were the only values in knowledge, operationism would have won a clear-cut victory. But we want our knowledge also to increase in extent, and this requires us to employ, as instruments of prediction, ideas which are vague and propositions which are uncertain. The conflict has been described in different terms among writers in the field, and has given rise to various proposed solutions. Operationally defined concepts are called "constructs"; all others "hypotheses." Or the former are called "intervening variables" and the latter "hypothetical constructs." Some would argue that only the former have a place in science. Others claim that the latter represent "traditional meanings" of the concepts and are justifiable on the basis of certain "intuitive methods" commonly employed in science. Still others give hypotheses a legitimate status in primitive science but exclude them

from mature science. Many writers give them only
"heuristic" value, arguing that they direct scientific activity but are ultimately to be abandoned since they constitute no part of the subject matter of the investigation.
All of these solutions are based upon the pragmatic fact
that what we want in knowledge varies from time to time
and from situation to situation; when we want clarity
and certainty we employ operational methods, but when
we want to extend the area of knowledge we supplement
or replace these methods by the logically less rigorous
procedures which are based on analogical inference,
hunch, intuition, and "extrapolation" beyond what is
directly given.

Once we have recognized this conflict we can see that
the knowing activity is shot through, at a lower level,
with similar oppositions—that between simplicity and
complexity, that between generality and singularity, that
between abstract thinking and thinking in "models," that
between descriptive procedures and "fictional" activities
involving the use of ideal types and other "as-if" notions,
that between constancy and flexibility in the meanings
of our symbols, and so on. Take the conflict between
simplicity and complexity as an illustration. The traditional textbook illustration is the opposition between the
Ptolemaic and Copernican theories of the heavens; and
the statement is usually made that each accounts for all
data but that the latter is preferred to the former merely
on grounds of greater simplicity. No doubt the search
for simplicity is an important motivating factor in science;
we can see it in the preference for "smooth" curves, in the
disregard of "disturbing" influences, and even in the very
use of proper names for the purpose of tying together
associated events or temporal series of events having a

certain continuity. But simplicity is not an unmixed blessing. Whitehead advises us to seek simplicity and then distrust it. And we are continually warned against over-simplification. The fact seems to be that at certain times and for certain purposes a simplified explanation is just what we want, but on other occasions we consider such information artificial and demand something which more accurately reveals the real complexities of the situation, even though our knowledge becomes less neat and less manageable.

Or take the conflict between descriptive procedures and those involving ideal types: One need not argue the desirability of descriptive adequacy in science. But the important role of "fictions," "conceptual shorthand" and other "as-if" devices (86:96) should not be overlooked. Their use clearly implies that at times we prefer not to describe the world, or even to employ hypothetical entities whose existence is doubtful, but actually to introduce into science entities which are believed not to exist. We state their behavior in contrary-to-fact conditionals. They are special cases of what are sometimes called "dispositional concepts." Through them we learn something about nature. For example, we gain information about a certain gas if we know how it would behave under conditions of temperature or pressure not yet attainable, or we learn something about a certain individual if we know how he would act under conditions, say, of extreme good or bad fortune, never yet actualized in his case. Clearly the introduction of such notions involves an abandonment of operationism in the narrow sense, since no such entities can be created by physical operations. But it also requires giving up the imperative that science be above all things clear and certain. "As-if" notions cannot possibly be as clear as concepts which are descriptively definable, since any properties attributed to them in contrary-to-fact conditionals will be based on such crude methods as analogy and serial extrapolation. Yet symbols of this kind have a demonstrated utility in science, in spite of the fact that they neither describe nor do they explain in the sense that verifiable predictions may be drawn from them.

5. Criteria for Adequate Knowledge

Our conclusions about knowledge may now be stated. Knowledge is not a simple thing, definable by a single quality or by a small number of mutually compatible qualities. Since this is true the usual formulations of the ideal of knowledge in terms of correspondence *or* of coherence *or* of workability are not adequate. Knowledge might possess all of these characteristics to the extent to which they are not incompatible. And to the extent to which they are incompatible we must be provided with a principle which will enable us to select that goal which is to dominate. This seems to make workability the final criterion, and there is probably some sense in which this is the case. But unless the kinds of workability are carefully distinguished, and unless the relations of these various modes of utility to some of the other ideals of knowledge are stated we shall have accomplished nothing. To say that knowledge is anything that enables us to solve problems and thus to restore the interrupted flow of behavior is a gross over-simplification; in fact it is essentially a tautology.

We have preferred, therefore, to formulate the criteria of adequate knowledge in terms of certainty, clarity and range (or scope), with the choice between these, in case there is a conflict, determined by a somewhat vague and

fluctuating criterion of workability. Two of these criteria
—certainty and clarity—are fairly general and are probably involved to a degree in all knowledge. But, as we
have seen, they must often be sacrificed, at least partially,
for the sake of predictability (capacity for referring beyond what has been experienced, capacity for integration). In essentially the same way, when we attempt to
attain one of these general goals, we must often make a
choice between other opposed kinds of knowledge—simple
and complex, singular and general, abstract and pictorial,
descriptive and ideal, constant and flexible. Certain of
these kinds are interrelated in complicated ways. For
example, since we can have no generic images strictly
pictorial knowledge must always be singular; again, general knowledge and abstract knowledge both involve some
loss in sensuous clarity; clear knowledge must be fairly
constant, and vague knowledge permits a certain flexibility; singular knowledge in its extreme forms permits no
necessary prediction since there is no deductive inference
from individual cases, and, conversely, a high degree of
logical integration involves only the most abstract descriptive adequacy. To argue that only one of these kinds is
"allowed" is to disregard the fact that they are all found
in science as a going enterprise. To argue that one kind
is "preferred" to any other requires a further indication of
how this preference is determined by the general nature
of the science and by the specific nature of the problem.

The result of these considerations concerning the nature
of knowledge is to place the problem of operationism in a
context which offers some possibility of solution. Whether
or not we shall employ operations, and if we do employ
them what they are to be, are determined by what characteristics we want knowledge to possess. If operations

are needed at all in knowledge, they are required as means, not as ends. And if they are required as means they must be selected with reference to ends. And, finally, if the ends are various and competing, the operations will be also. Our task, then, will be, first, to decide whether operations are needed in the production of knowledge, and, second, on the basis of an affirmative answer to this question, to list the main kinds of operation found in actual knowing situations and to define each of these kinds in such a way as to indicate so far as possible its exact role in knowledge. This is a large order and can be filled only in barest outline.

6. Role of Operations in Knowledge

The first question—whether we need operations at all in the acquisition of knowledge—can be answered by a verbal decision. *Knowledge* is the end result of an *activity* performed upon *something-which-is-to-be-known*. These three, in their proper relationship, and with the possible addition of the knower, who functions as the director of the activity and as the residence of the knowledge, constitute what may be called the "cognitive situation." But once we have recognized that knowledge never exists except as a result of an activity of knowing we are confronted with a terminological problem: Shall we use the word "operation" as synonymous with any knowing activity, or shall we restrict it to certain of these activities? If we choose the former we shall make non-operational thinking impossible by definition. But this will provide little comfort to most of the operationists since we shall be using the term in a highly general sense which they would probably reject. This is, in fact, what Bridgman does in his later writings where he identifies operations with any activity involved in the attempt to reduce happenings to understandability (*21*:115). If we choose the second alternative the consequences will be equally definite. We shall be under obligation to indicate exactly what knowing activities are to be characterized as operational, but also to admit that non-operational activities are legitimate to the extent to which they eventuate in knowledge.

We shall take the former alternative. An operation may then be identified with any act which is performed with a view to the production of symbolic knowledge, or to its improvement in clarity, certainty, extent, or any other of the more specific ideals mentioned above. All knowing is operational, and all operations are functionally determined by the kind of knowledge they are designed to produce. Restriction of knowledge to that which is symbolic creates some problems which will be handled later. For example, discrimination is a cognitive operation, yet one cannot easily determine whether it is also symbolic.

7. Two Meanings of "Functional"

The use of the term "functional," however, may produce precisely that confusion which an earlier discussion was designed to avoid. The concluding pages of Chapter V were devoted to an attempt to distinguish symbolic operations from those which are non-symbolic, and it was there pointed out that only symbolic operations are cognitively significant. Since we have now generalized the term "operation" to include any cognitive act, we must make a special effort to distinguish cognitive operations from those which are non-cognitive. The difficulty at this point is that there are two distinct meanings of the word "functional" involved.

Let us return to the example of taking the temperature of a gas by inserting a thermometer in it. This can be called a cause-and-effect situation, and in accordance with common usage the effect may be said to be a function of the cause in the sense that both the existence and the character of the effect are dependent on the existence and character of the cause. Furthermore, in this case part of the cause is an operation, *viz.*, inserting the thermometer in the gas and waiting a few moments. Now consider this situation from the point of view of the three kinds of operations mentioned in the earlier discussion. (1) There is a physical operation producing a physical event, and the event may be said to be a function of the operation. But the situation, as it stands, need not be cognitive. (2) Even if the reading on the thermometer is taken as a sign of the temperature of the gas in the same way that a red sky in the evening is taken as a sign of clear weather on the following day, the reading may be said to be a function of the operation but the situation would not usually be called cognitive in the strict sense. Animals, of course, exhibit this kind of behavior, and if it is to be called cognitive then animals may be said to think. But (3) the act of devising a symbol (say, employing a number) to represent the temperature of the gas would be cognitive in the strict sense. For here the symbol, as distinguished from the sign, is created by the knower and given a place in a language (64:35-6). In this case also the symbol may be said to be a function of the operation, since its meaning is determined both by the operation and by that on which the operation is performed. Thus one might set up the general formula

$$y = f(x)$$

where y is any effect and x is any cause (including in the term "cause" both the operation and that on which the operation is performed). Then y, the position of the mercury in the tube, might be a mere effect of x, or it might also be a sign of x; but only as a certain number, *i.e.*, as an element of a mathematical language, could y be a *symbol* of x, and hence cognitively refer to x. But in all these cases y would be a function of x.

However, there is also involved here another meaning of the term "functional." Anything is functional if it contributes to the realization of some end; in this sense anything which is instrumental is functional. According to this meaning one ought to say not that y is a function of x, but that x is a function of y, and the mathematical representation of the relationship would be completely misleading. Inserting the thermometer in the gas is functional in so far as it is used to produce: (1) a reading on the thermometer as a physical event; (2) a reading on the thermometer as a sign of the temperature of the gas, or (3) a numerical symbol which is a certain kind of knowledge about the gas. Only in the last of these three senses is the operation *cognitively* functional. The operations to be discussed below are all cognitively functional; they are designed to produce not mere effects, or even signs, but symbols which will be elements of knowledge systems and which will have their adequacy measured in terms of the various ideals sought in such knowledge systems. Cognitive operations are selective acts chosen with reference to the creation of knowledge.

We may say, therefore, that cognitive situations are functional: (a) in the sense that the symbols employed are *defined* by operations and by things operated upon, and (b) in the sense that the operations are *selected* with a view to producing the desired kinds of symbols. The first

meaning of "functional" indicates the sense in which cognition is *causal*; the second indicates the sense in which
it is *teleological*.

8. Three "Elements" of the Cognitive Situation

Before attempting a classification of the main types of
operation one further point should be clarified. In view of
the criticisms of operationism in the previous chapters certain general features of the role of operations may now be
stated.

We may restate the functional formula given above,
applying it specifically to the cognitive situation, as follows:

$$S = f(D, O)$$

In this formula "S" represents any symbol, "D" represents
any datum, "O" represents any cognitive operation; and
the formula states that the meaning of any symbol is a
function, in the first sense of this word, of certain things
which are given to be known and certain cognitive operations performed on these data.

From this formula, as well as from previous considerations, we can readily see what would happen to the cognitive situation if any one of these elements, S, D, or O, were
to be eliminated. Suppose, first, that we were to drop D
and say that the meaning of a symbol is a function only of
operations. This is substantially what would be meant by
anyone who said that a concept is *synonymous* with a set
of operations, and its absurdity should now be apparent.
Operations must always be performed on something;
otherwise knowledge would be produced only by knowing, not by knowing *anything*. Suppose, secondly, that
we were to drop O and say that the meaning of a symbol
is a function only of data. This would also reduce to an

absurdity; it would make all predictive knowledge strictly impossible. It would be to abandon not only operationism, but any activistic conception of knowledge. Knowledge is clearly the result of a process; it does not just happen but is built up by action and effort, and its nature is determined as much by *how* we know as by *what* we know. The existence of "-ing" forms—inferring, symbolizing, naming, generalizing, inducing, deducing, hypothesizing, describing, measuring—all testify to this fact. Suppose, finally, that we were to drop S. The same kind of difficulties would arise, for we should then have something to be known and a method of knowing but never any knowledge which is the outcome; we should have potential knowledge, not actual knowledge. But potential knowledge is certainly distinct from actual knowledge, and each of these is distinct from the process of knowing. The need for distinguishing between knowledge and that which is only capable of being known, is indicated by the persistence in the history of epistemology of some element of potentiality in the cognitive situation. Illustrations can be found in Locke's substance, Kant's thing-in-itself, Mill's Permanent Possibility of Sensation, the logical positivists' dispositional predicates (malleability, conductivity, solubility), Lundberg's "that which evokes the responses" (59:9), and Feigl's conception of things as "know*able*" (86:257). All of these create the paradox of "saying something," and therefore providing actual knowledge, about a world which we know only as potentiality; yet they—or similar notions—are required if the cognitive situation is to be understood even in its minimal features. Without something to be known, the knowing, and the knowledge, the cognitive situation simply does not make sense.

9. Classification of Operations

Some preliminary remarks about the following classification of operations should now be made. It will not include physical operations, directed either at the production of mere effects or at the production of signs. This is not to say that physical operations may not accompany most of the cognitive operations which are to be listed; discrimination, for example, requires all of the physiological operations used in the effective functioning of the sense organs, and even "thinking" itself would presumably be impossible without brain activity. Measurement also usually employs manipulatory acts, and classificatory and ordering operations are frequently accompanied by actual physical groupings of the objects to be classified, as in the arrangement of museum displays. All of these physical operations will be disregarded, but not denied, in the treatment which follows.

Furthermore, the list is presumed to be suggestive rather than exhaustive, and its items are not supposed to be mutually exclusive. To construct an adequate list of cognitive operations would be to solve the problem of knowledge. What is given below is merely a proposed classification of the most important kinds of cognitive operations. In view of the limitations of space only a very brief characterization of each operation can be given.

10. Discriminating

Probably the most basic of operations is that of discriminating; it may also be called "inspecting," "attending to," "discovering" and "becoming aware of." The words "introspecting" and "perceiving" should be avoided since they tend to introduce the subjectivism-objectivism controversy, which is irrelevant at this point. The word "in-tuiting" has the required flavor, but it is used in so many senses that its employment is not to be encouraged. Russell's "knowledge by acquaintance" as opposed to "knowledge by description" is approximately what is in-tended. The element of directness or immediacy, ex-pressed by saying that there is nothing that "comes be-tween" the knower and that which is known, seems essential to a characterization of the operation.

The main problem in clarifying this operation is to de-termine whether it ever occurs in its "pure" form, or whether there is also involved in it at least a naming opera-tion (employing a strict proper name), and possibly even a generalizing operation (employing a concept or a com-mon name). The problem can also be formulated by asking whether discrimination is a cognitive operation. In view of the definition given above, in order to be so it must involve the creation of symbols. Hence if it is a case of a mere awareness without even such words as "this" and "that" to identify what has been presented it cannot be called a cognitive operation.

One possible solution to his problem, which will here be stated dogmatically, is that discrimination and pure nam-ing are one and the same thing. What this means is that in becoming aware of any particular there is always an actual or incipient naming act involved; this is essentially a "tagging" act by which we create a demonstrative symbol (the best examples are "this," "that," "here," "now" and "I"), which then refers back for its meaning to the thing discriminated. Discrimination, however, is not the same thing as the generalizing operation employed in the estab-lishment of concepts, but is operationally distinguishable from it; to become aware as a particular of that which is being discriminated is not the same act as to become aware of it as representing a kind. Finally, whether discrimination is or is not separable from generalization, *i.e.*, whether we ever discriminate (name) without also at the same time generalizing, is an important question but one whose answer is not required for our discussion. This may all be summarized by saying that discrimination is *indistinguishable* from pure proper naming and *distinguishable* from generalization; but whether discrimination ever *occurs apart from* generalization is undetermined.

This permits us to define discrimination as the operation by which pure proper names are created and given meaning. In terms of our basic formula, the meaning of a pure proper name is determined both by the datum of which we are aware and by the discriminatory operation. But since the act of awareness is presumed to be "external" to the thing which is being discriminated, *i.e.*, neither to modify nor to create it, the meaning of a pure proper name is determined almost wholly by the thing which is being named. Hence pure proper names in the presence of their referents have the possibility of the maximum clarity among all symbols which we employ, and in the absence of their referents are meaningless. Confusion should be avoided, of course, between the pure proper names being considered here, and the more common, or impure, proper names which function as such in actual languages; the latter, as we shall see in a moment, play the role of abbreviated descriptions, since they require that their referents have a certain qualitative identity in time. Pure proper names are required to provide a place in knowledge for that kind of act which can be expressed by saying, "I am now directly aware of such-and-such"; without an act of this sort knowledge would be impossible.

11. Associating

The second type of operation may be called "associating." It is that by which names for "things" are devised and given meaning, and it is distinguished from discrimination in that it involves combining particulars, for the purposes of symbolic representation, into complex and enduring objects. Any such object then becomes "that which" is "constituted by," or "contains," the particulars, both those which now coexist and those which make up the past and future life of the object.

There are two main kinds of association—by coexistence and by succession. The operation of association by coexistence involves grouping particulars into complexes by virtue of the fact that they occupy the same, or approximately the same, space and time; for example, a certain shape, size, color, texture, weight and impenetrability are, by virtue of their "togetherness," combined into a table. Frequently this association is based also on the dependence of certain members of the complex on others; for example, if the size of the table were to be changed the weight might also change, or if its weight as measured by a coil spring were to be changed the weight as measured by a balance might also change. It is this fact of association which the operationists have vaguely in mind when they recognize that we can "define the same concept by different operational routes" (Feigl, 86:255), or that two different operations are equivalent (Bridgman, 21:121; 86:247-8), or that their results show a high coefficient of correlation (Chapin, 35:ch.XIX).

The operation of association by succession involves grouping particulars into enduring objects by virtue of certain resemblances which persist through time and by virtue of causal relationships which hold among particulars. The impure proper names of language, as well as what are commonly called "descriptive phrases," are given operational meaning by a combination of association by coexistence and association by succession. For example, by "Eisenhower" we mean not simply a certain complex of particulars which could be seen by going to Washington, but another complex of past particulars which constitutes Eisenhower's earlier life, and still another complex which makes up his future activity. These stages in his life are associated both by similarity, since he remains more or less constant in appearance, personality and political outlook, and by causal connection, since what he is today is the result of what he was in the past, and the cause of what he will be in the future.

Now unless these associative operations are defined as clearly as possible we get into difficulties of the kind which arose in our critique of operationism, but which could not be adequately discussed at that point in the absence of a conception as to the goal of knowledge. Bridgman formulated the problem in terms of a conflict between the desire to be clear and precise, even though this require an extensive vocabulary, and the desire to use a small number of words, with a corresponding increase in vagueness and ambiguity. We can now formulate it as a conflict between the clarity of our symbols and the simplicity of our symbolic scheme. To meet the former goal with complete adequacy a vocabulary consisting wholly of pure proper names would be required; but there would need to be a very large number of them, for each particular as it occurs would require a special name. To meet the latter goal with complete adequacy a language consisting of a small number of proper names would suffice; but the names would not be "pure" since we could not specify exactly what complex of coexistent and sequential particulars would be included in the referent of any name. As Bridgman says, common usage prefers ambiguity (or vagueness) and a small number of words to precision and a large number of words. "Eisenhower," for example, is vague in two ways. It is vague in the actual scope of coexisting qualities which we now mean by this name; for example, it refers to certain qualities which we can detect, such as height, weight and general appearance; but it also refers to certain qualities which we cannot detect, such as the condition of his internal organs, his score on an intelligence test (in case he has never taken one) and what he is thinking about at some moment when he is not speaking or acting. Thus we cannot say clearly what we mean by the name "Eisenhower—now." But the name is vague also in its reference to past and future; we use it to designate all he has done and experienced in the past, some of which may be known but much of which is not known even to Eisenhower himself; and we use it also to designate all that he will do in the future, very little of which anyone knows.

We can therefore see that in the introduction of associational operations which define the usual proper names we must sacrifice a certain amount of clarity, but our symbolic system gains in simplicity. To take the stock example, if "the intelligence of Mr. A" is defined in terms of the results of some particular test, and only in terms of these results, certain consequences follow: the word is a pure proper name, clear in its definition, but it cannot later be broadened in scope to include other measures of the intelligence of Mr. A, for there can be no such thing. On the other hand, if "the intelligence of Mr.

A" is defined in terms of the result of some particular test *plus* the results of certain tests to be devised in the future, all of which we characterize in terms of Mr. A's mental alertness, logical acumen or discernment, then other results follow: the word is a proper name in the ordinary sense, vague in its definition, but not requiring redefinition when later tests are invented. It cannot, of course, be a common name, since no generalizing operation has yet been performed.

Stevens (83:519) seems to me to be confused on this point. He argues that no concept should ever be allowed to congeal and that operational procedure insures against fixation. And he then points out that the operational basis of a concept may be "added to and the concept renamed." What is really the case, however, is that operational procedures, if interpreted narrowly, do not insure against fixation; they *guarantee* it. A pure proper name, once defined, cannot be redefined; hence if such a name is once given meaning in terms of a discriminatory operation its content is fixed for all time. But if we admit associational operations as well as discriminatory operations the names which we invent become sufficiently vague to permit later broadening of meaning without actual redefinition. This is only to recognize that one of the many conflicts in the goals of knowledge has come into the picture—the conflict between rigidity and flexibility. For certain purposes we want a fixed language; it aids us in communicating with one another, in using the accumulated wisdom of the past, and in achieving clarity and precision. But it requires a multiplication of symbols, for a particular in *my* experience may not be identifiable with one in *yours,* and what *we* experience today may not be precisely what *our ancestors* experienced; hence we must continually invent new symbols to refer to these new data. This suggests that a more flexible language, one which is capable of adapting itself to what is common at once to my experience, your experience, the experience of the past, and what may be disclosed in the experience of the future, has certain advantages. It is vague, but it allows for novelty without an endless addition of words to our vocabulary.

There are, of course, many other kinds of associational operations, determined by the various kinds of "wholes"—mechanical, organic, social, political, psychological, artistic and so on—which we believe to exist. Furthermore, just as there are associational operations by which we combine elements into wholes, so there are analytic operations by which we break up wholes into elements. Many authorities, for example, would insist that what is disclosed by the basic discriminatory operation is not *sense-data* (or particulars, in the strict sense), which we then combine into *objects* by *associational operations*. On the contrary, what we directly perceive is *objects* which we then *analyze* into *sense-data*. The issue involved is too complicated to be discussed in this short monograph. All that need be admitted here is that the basic discriminatory operation is neither associational nor analytic, but that what is disclosed through this operation may be further operated upon by associational and/or analytic operations.

12. Generalizing

The third type of operation is that of generalizing. This may be considered, if desired, as a special type of associational operation; for by means of it certain types of wholes (classes) may be considered to be formed by association through resemblance. "Generalizing" will here be defined broadly so as to include the formation of concepts or general names (classifying, abstracting, universalizing), and the formation of general laws (induction). While these two processes are logically quite distinct, since the former starts with particulars and creates terms, and the latter starts with associations of particulars and creates propositions, nevertheless they both involve the "leap of thought" by which symbols are created having reference beyond the actual cases examined. In the case of concepts this produces an inevitable vagueness in the symbols; we cannot mean by any general name simply the particulars which led us to invent the name in the first place (Russell says (79:158) that by "hot" we mean whatever causes us to utter the word "hot"), since in this case the name would be a pure proper name, not a general name at all. By the denotation of a general name we mean such-and-such actual particulars plus an indefinite number of others bearing to these particulars a degree of similarity which is not at the moment specified. This produces vagueness very much like that involved in the usual proper names, and creates the same difficulties. If we attempt to define "red" with a high degree of precision, including in its referents three given spots of red, r_1, r_2, and r_3, we cannot use the same name to characterize r_4 but must invent a new one; on the other hand, if we allow an element of vagueness in our definition, we can use the same name to cover the new case. Again we are confronted with a choice between a language which is clear but contains many words, and one which is vague but not so complex. The same difficulty exhibits itself in the inductive problem where we have a choice between the method of simple enumeration, which does

not result in laws at all and where there is no reference
to unobserved cases, and the genuinely inductive procedure, which does result in laws but covers unobserved
cases only in a problematic manner. Here generality and
certainty are in conflict.

13. Ordering

The fourth type of operation may be called "ordering";
it is that by which symbols for series are devised and
given meaning. An ordering operation is one by which a
symbol is defined to represent particulars which are
united into a complex by an asymmetrical, transitive and
connected relation. These properties of an ordering relation are defined (77:29-41) as follows: A relation is
asymmetrical if when it holds between x and y it does not
also hold between y and x. A relation is transitive if when
it holds between x and y, and between y and z, it then
holds between x and z. A relation is connected with reference to a class which is to be ordered if either it or its
converse holds between every pair of elements of the
class. Thus ordering, like generalizing, may be considered
as a kind of associational operation, for it results in the
creation of a certain type of whole. Such an operation
is employed in the definition of a proper name of the
usual type, since the events making up the life of an individual may be arranged in temporal and causal orders.
But there are other temporal and spatial relations which
determine orders; for example, Whitehead's relation of
extension, by which events contain or include other
events, and the spatial relations *to-the-left-of* and *above,*
which determine orders under certain conditions. Other
ordering relations are neither spatial nor temporal, such

as the relation of *heavier-than* holding among weights, or the relation *preferable-to,* holding among foods or aesthetic objects. Within this general framework of ordering relations there arise distinctions between relations which generate quantitative and those which generate non-quantitative (comparative) orders, and between those which generate intensive and those which produce extensive quantitative orders (33:8-15; 49:54-78; 78:Pt. III; 32:Pt. II; 85:21-30). Detailed discussion of these types is unnecessary here. All that is required for our purposes is to recognize that symbols for ordered particulars are defined by specifying certain ordering operations, and that these are in turn based on discovered asymmetrical, transitive and connected relations holding between the particulars.

Whether symbols for orders are pure proper names, impure proper names or generalized symbols will depend upon the nature of the operations employed in their derivations. A symbol for an order will approach the status of a pure proper name to the extent to which it is designed to refer only to the collection of particulars among which the asymmetric, transitive and connected relation holds. It will be an impure proper name if it is designed to refer to this collection plus certain other particulars, not at the moment specified, but assumed or to be added later when their relations to the collection are discovered. It will be a generalized symbol if it is designed to refer to other collections exhibiting the same kind of ordinal principle.

14. Measuring

The importance of order lies in the fact that measurement depends upon it. The relation between ordering and measuring may be stated in an over-simple manner as follows: A series of particulars is metricized if there exists a procedure by which it may be correlated with the series of real numbers in the usual arrangement. This procedure then permits a number to be "attached" to each particular and to constitute its measured value. A measured value is called "fundamental" if there is no other measured value on which it depends. A series which permits fundamental measurement possesses the property of being "additive," *i.e.*, there exists a specific mode of combining any two elements in the series into a new element which may be considered as the sum of the combined elements; space, time and mass are examples of fundamental measurements. A measured value is called "derived" if there exists a procedure by which it may be obtained from another measured value or other measured values; for example, temperature is measured by the length of a column of mercury, density is measured by dividing mass by volume, and velocity by dividing space by time.

Hempel (49:sec. 13) points out that derived measurement exhibits itself in two forms, by stipulation and by law. Derived measurement by stipulation involves the invention of a "new" quantity which is defined as a certain mathematical function of other quantities already known; for example, all averages are of this kind (velocity, population per square mile, annual income, time required for an animal to get out of a maze, I.Q. of a group of individuals). But since the stipulation is presumably *free*, any mathematical function of other measured values could define a derived measurement; thus the square root of the product of a man's age and his height would constitute a legitimate example of a stip-

ulated measurement, as would also the quotient of his
I. Q. and his basal metabolic rate, the logarithms of the
size of his shoes as he grows older, and the example given
by Adler in Chap. V. All that is required is that the
measured values employed in the formula be precisely
defined, and that the mathematical function of these
values be accurately specified.

Derived measurement by law does not introduce a
"new" quantity but merely an alternative way of measuring one that has been previously introduced. This
is done by discovering some law which represents the
magnitude in question as a mathematical function of
other quantities, for which methods of measurement have
previously been laid down (49:70). Examples are the
measurement of altitude by the barometer, of specific
gravity by the hydrometer, and of distance by the trigo-
nometric method.

What, then, does a metrical symbol mean? It means
one of three things, depending upon whether it expresses:
(1) a fundamental measurement, (2) a derived measurement by stipulation, or (3) a derived measurement
by law. (1) To say that a line has a certain length is
to say that it is a member of a collection of lines which
(*a*) can be ordered by an asymmetrical, transitive and connected relation; (*b*) possesses "additive" properties; and
(*c*) may be correlated by means of a physical procedure
with the series of real numbers in the usual arrangement.
(2) To say that a gas has a certain density means simply
that the measured value of the mass has been divided
by the measured value of the volume, and a new number obtained. (3) To say that a mountain has a certain
height as measured by a barometer means that the mountain is a member of a series of objects whose heights as

measured by tapelines exhibit a certain degree of correlation with their heights as measured by atmospheric pressure, and hence that measurement in terms of the latter may be substituted for measurement in terms of the former. In the notion of stipulation there is no requirement that the derived measurement have any empirical correlation with any fundamental measurement, or with any other derived measurement. For example, the measurement of density by the quotient of the mass and the volume is an operationally satisfactory definition of this quantity, independently of the fact that liquids arranged according to such a principle exhibit a high degree of correlation with liquids arranged according to the serial principle *floats-in*. But stipulated measurements are employed in the hope that they may later be discovered to have empirical correlations with other measurements. Similarly, in the notion of measurement by law there is no requirement that the correlation be expressible by means of a simple mathematical function. Derived measurements by law are often brute correlations which, it is to be hoped, will later be rendered mathematical by a properly formulated theory.

Since quantitative symbols presuppose symbols for orders, the former are no more precise than the latter, and they are of essentially the same kind. To take the second point first, quantitative symbols are either pure proper names, impure proper names, or concepts, depending on whether they are derived by discriminatory operations only; by these together with associational operations; or by discriminatory, associational and generalizing operations. A symbol for a metricized series approaches the status of a pure proper name to the extent to which it is designed to identify an actual collection of particulars exhibiting the proper serial relations; in this capacity it plays essentially the same role as the word "this." A quantitative symbol is an impure proper name if it designates such a collection as extended by interpolation or extrapolation to include other particulars to be discovered later; in this capacity it plays essentially the same role as the word "Eisenhower." A quantitative symbol is a generalized symbol if it refers to such a collection as extended to include other collections bearing a certain degree of resemblance to the given one; in this capacity it plays the same role as any concept, *i.e.*, it refers to a class of instances.

With reference to clarity, quantitative symbols are as clear as their operational derivations permit. Pure proper names have the maximum of clarity. Among metrical symbols those for fundamental quantities and those for derived quantities by stipulation have the greatest clarity; to the extent to which the processes of fundamental measurement and the functional relationships of the derived quantities to the fundamental quantities are accurately specified, the symbols are clear. Symbols for derived quantities by law, being obtained from the data by indirect operations, are correspondingly less clear.

15. Analogizing

The fifth type of operation is that of analogizing. It involves the creation of symbols having more or less resemblance to actual particulars. The operation always explicitly or implicitly involves generalization, since noting of resemblances in going from actual particulars either to symbolic ones or to symbolic *kinds,* is always a part of the operation. Furthermore, since we never wish to create exact duplicates, disanalogizing operations accompany the analogizing operations. By means of such complex operations symbols for entities more or less like actual entities are created.

The importance of analogizing operations lies in the fact that by means of them the usual hypotheses, theories and constructs of science are derived and given content. In Chapter II we found that these concepts, called "postulational" by Northrup, fall roughly into two classes, inferred concepts and constructed concepts, and that both of these are to be distinguished from descriptive or intuitive concepts, which refer directly to data. Postulational concepts are characterized by the fact that their meanings are not determined wholly by the data or intuited concepts which originated them. Since we cannot define them merely by inspecting certain data we must do it operationally by some sort of act performed on entities which are directly given. This produces, on the one hand, constructs, which have only a "linguistic status" and refer merely to the data which originated the operations; and, on the other, inferred concepts, which are presumed to refer not only to the original data but also to certain entities having a hypothetical existential status. Analogical operations may be either constructive or inferential. The concept of the perfect lever, for example, is derived from actual levers by constructive, analogical (and disanalogical) operations, and is commonly presumed to have only a "fictional" status. The concept of the wave motion of light, on the other hand, is derived (or could have been derived) from the wave motion of sound by inferential, analogical (and disanalogical) operations, and is commonly taken as an hypothesis having a certain degree of confirmation.

Analogizing operations (with their associated dis-analogizing operations) are of many kinds. Sometimes they start with concrete particulars and end with particulars which are equally concrete but have been produced by modifications of those from which they originated. Frequently such concrete analogizing operations have physical counterparts in the creation of actual models made of string, rubber, pulleys, wire, *etc.*; examples are to be found in museum models of the astronomical system, of the structure of the atom, of the nervous system, and the like. Certain scientists, notably Faraday, Kelvin, Lodge and Maxwell, have argued strongly for this type of symbol in science (*41:* Ch. IV); Lodge even insisted that one does not truly understand any phenomenon unless he can construct such a model. Much more common than actual models are those which fall into the general class of pictures, diagrams, sketches, maps and outlines. Peirce (*69:*2.276, 2.277) includes images and metaphors as well, and suggests the word "icon" to characterize all devices which represent by means of similarity. Peirce even includes mathematical symbols as a type of icon. Bridgman (*14:*52-3) argues that the model is a useful and, in fact, an inescapable tool of thought, since it enables us to think about the unfamiliar in terms of the familiar.

Very common among analogical operations are those employing serial extrapolation. These are the operations by which symbols are derived representing extreme cases of variables known to exhibit themselves only within a limited range. Eddington calls extrapolation the method of "like this only more so" (*42:*247). By such an operation we are able to create symbols representing quantitative values beyond any which are descriptively known. Sometimes these involve merely "next" cases—slightly higher velocities, or greater densities, or shorter durations. But much more important are the so-called "limiting" or "perfect" cases obtained by arranging actual particulars serially and then extending the order by progressively eliminating "disturbing influences" until a limiting case has been reached. Physical examples are infinite and zero velocities, masses concentrated at mathematical points, perfect levers, frictionless motions and ideal gases. Examples from the social sciences are the usual ideal types—isolated individuals, purely economic men, and utopias.

16. General Conclusions

In view of our previous discussion of operations we can now see that there is no sound reason for restricting the use of the term "postulational symbol" ("hypothesis," "theory," "construct") to the results of analogical operations. All symbols except pure proper names have elements of vagueness and conjecture; the usual proper names and descriptive phrases include reference to undesignated associated qualities, and undesignated past and future particulars; generalized symbols include reference to unobserved particulars and correlations of particulars determined by unspecified resemblance to those which are observed; symbols for orders and symbols for quantities have the same kind of vagueness. In this way every operation except pure discrimination introduces vagueness and conjecture, and every symbol except the pure proper name introduces something hypothetical. As a consequence, if the operationist insists on *perfect* clarity his language is reduced to a series of such words as "this" and "that," and his activity to a series of gestures of pointing. But if he is willing to grant that knowing is more than mere discerning and involves anticipating what nature will later disclose, he can permit himself a more extended language and a more varied series of cognitive acts, but he is then committed to an element of vagueness and uncertainty. This can, of course, be reduced by more accurate specification of the operations performed—associating, generalizing, ordering, measuring and analogizing—but in no case can the clarity of pure proper names be achieved. This is simply another formulation of the conflict, referred to earlier in the chapter, between certainty and clarity, on the one hand, and predictability, on the other. We can now express it by saying that if we are content with what we already know—facts—we can be both very clear and very certain, but if we wish to know something else—hypotheses—we must sacrifice to a degree both clarity and certainty.

This brings into clear focus the essence of the operational theory. We want our symbols to have the highest degree of clarity and certainty. As strict empiricists we recognize that this can best be achieved if we restrict ourselves to symbols whose meanings are determined wholly by the ostensive method. But we soon realize that this gives a very much impoverished language, and provides us with no techniques for extending the area of what we know. So we introduce operations. These seem to solve our problem, for they provide us with methods for introducing new concepts, yet concepts which need not be vague if we are careful to define our operations precisely. But we then realize that among the many operations which we may employ only that of discrimination permits the necessary clarity, and this provides us with a language merely of pure proper names. Such a language, unfortunately, is no better than the highly impoverished one we started out with as strict empiricists. To avoid this limitation we may make concessions, reluctantly perhaps, adding first only associational operations. But while we suffer only a slight loss in clarity as a result of this addition, we achieve only a small gain in predictability. If we progressively add generalization, ordering, measuring and analogizing we lose in clarity and certainty but gain in predictability. (Incidentally, if, as we include these later operations, we abandon discrimination and association, we then become progressively more muddle-headed and speculative.) Whether we accept all or only some of the operations will depend, in general, on our view of the goal of knowledge, and, in particular, on the demands of the specific problem at hand. A thorough grasp of the nature of knowledge discloses the fact that this is not the only opposition found in cognitive activity, but that the conflicts between simplicity and complexity, rigidity and flexibility, and the others listed earlier in the chapter, call for similar decisions.

Hempel (49:28-9), in calling attention to the partial indeterminacy of meaning in certain concepts, suggests that this be called their "openness of meaning" (89). This is a very fortunate term, and I propose in this concluding paragraph to employ it, but to use it in a manner which is probably much more general than Hempel would be willing to accept. I suggest that all symbols be characterized in terms of the degree to which they are open. Pure proper names are closed; they are defined in terms of the simplest and most basic of operations; and they have the maximum in clarity, in definiteness of reference and in existential commitment. As elements of symbolic schemes they are, therefore, indispensable. But if knowledge is to grow, a less narrowly conceived operationism is required; symbols of various degrees of openness are demanded. This concession does not mean an abandonment of the operational point of view since open symbols can also be operationally defined. But it does mean that the operations employed in their definitions, even when themselves carefully defined, permit, again in various degrees, vagueness, generality, predictability, indefiniteness of reference, uncertainty of existential commitment and flexibility. Whether these qualities are or are not desirable will depend both on one's conception of the goal of knowledge and on one's formulation of the specific problem at hand. But that open symbols are as indispensable in knowledge as proper names seems certain. The task of an operational theory is therefore one of listing and defining as clearly as possible the various operations which enter into the cognitive situation. The present chapter has attempted this in outline only.

BIBLIOGRAPHY

1. Adler, Franz: Operational definitions in sociology. *Am. J. Sociol.*, 52:438-44, 1947.
2. Allport, Gordon W.: The psychologist's frame of reference. *Psy. Bull.*, 37:1-27, 1950.
3. ———, and Bruner, J. S.: Fifty years of change in American psychology. *Psy. Bull.*, 37:757-75, 1940.
4. Alpert, Harry: Operational definitions in sociology. *Am. Sociol. Rev.*, 3:855-61, 1938.
5. Ballard, E. G.: Operational definitions and theory of measurement. *Methodos*, V:233-9, 1953, (with comments by P. W. Bridgman, 240-1).
6. Bavink, Bernhard: *The Natural Sciences.* New York, Century, 1932.
7. Benjamin, A. Cornelius: Operationism—a critical evaluation. *J. Philos.*, 47:439-44, 1936.
8. ———: *Introduction to the Philosophy of Science.* New York, Macmillan, 1937.
9. ———: The operational theory of meaning. *Philos. Rev.*, 16:644-9, 1937.
10. ———: The unholy alliance of positivism and operationism. *J. Philos.*, 39:617-25, 1942.
11. ———: Philosophy, the cult of unintelligibility. *Philos. Rev.*, 57:347-62, 1948.
12. Blumer, Herbert: The problem of the concept in social psychology. *Am. J. Sociol.*, 45:707-19, 1940.
13. Boas, George and Blumberg, A. E.: Some remarks in defense of the operational theory of meaning. *J. Philos.*, 28:544-50, 1931.
14. Bridgman, Percy W.: *Logic of Modern Physics.* New York, Macmillan, 1927. Courtesy of the publisher.
15. ———: The new vision of science. *Harper's*, 158:443-51, 1928-1929.
16. ———: Limitations of cosmical inquiries. *Sc. Monthly*, 37:385-97, 1933.
17. ———: A physicist's second reaction to mengenlehre. *Scrip. Math.*, 2:101-17; 221-34, 1934.
18. ———: *The Thermodynamics of Electrical Phenomena in Metals.* New York, Macmillan, 1934.

19. ———: *The Nature of Physical Theory*. Princeton, Princeton University, 1936.
20. ———: *The Intelligent Individual and Society*. New York, Macmillan, 1938. Courtesy of the publisher.
21. ———: Operational analysis. *Phil. Sc.*, 5:114-131, 1938. (Also included in *Reflections of a Physicist*.)
22. ———: Society and the intelligent physicist. *Am. Phy. Teacher*, 7:109-16, 1939.
23. ———: *The Nature of Thermodynamics*. Cambridge, Harvard University, 1941.
24. ———: Some general principles of operational analysis. *Psy. Rev.*, 52:246-9; 281-4, 1945. (Also included in *Reflections of a Physicist*.)
25. ———: Some implications of recent points of view in physics. *Rev. Int. de Phil.*, 3:479-501, 1949.
26. ———: *Reflections of a Physicist*. New York, Philos. Lib., 1950.
27. ———: The operational aspect of meaning. *Synthese*, 8:251-9, 1950-1951.
28. ———: Philosophical implications of physics. *Am. Acad. Arts & Sc.*, 3:1-6, 1950.
29. ———: The nature of some of our physical concepts. *Brit. Jn. Philos. Sc.*, 1:257-72; 2:25-36; 142-60, 1952.
30. ———: Remarks on the present state of operationalism. *Sc. Monthly*, 79:224-6, 1954.
31. Brunswik, Egon: *Conceptual Framework of Psychology*. Chicago, Univ. Chicago, 1952. (Int. Ency. of Unified Science, Vol. I, no. 10.)
32. Campbell, Norman: *Physics, the Elements*. Cambridge, Cambridge Univ., 1921.
33. Carnap, R.: *Logical Foundations of Probability*. Chicago, Univ. Chicago, 1950.
34. Cassirer, Ernst: *Substance and Function*. Chicago, Open Court, 1923.
35. Chapin, F. Stuart: *Contemporary American Institutions*. New York, Harper, 1935.
36. ———: Definition of definition of concepts. *Social Forces*, 18:154, 1939.
37. Dingle, Herbert: *Science and Human Experience*. London, Williams and Norgate, 1931.
38. Dodd, Stuart C.: A system of operationally defined concepts for sociology. *Am. Soc. Rev.*, 4:619-34, 1939.
39. ———: Operational definitions operationally defined. *Am. J. Soc.*, 48:482-9, 1942-1943.

40. ———: *The Dimensions of Society.* New York, Macmillan, 1942.
41. Duhem, Pierre: *La théorie physique.* Paris, Rivière, 1914.
42. Eddington, Arthur S.: *Nature of the Physical World.* New York, Macmillan, 1929.
43. Feigl, Herbert: Logical Empiricism, in *Readings in Philosophical Analysis.* New York, Appleton-Century-Crofts, 1949, 3-26.
44. ———: Logical analysis of the psycho-physical problem. *Philos. Sc., 1:*420-45, 1934.
45. Frank, Philipp: *Foundations of Physics.* Chicago, Univ. Chicago, 1946. (Int. Ency. of Unified Science, Vol. I, no. 7.)
46. ———: *Modern Science and its Philosophy.* Cambridge, Harvard Univ., 1949.
47. Hart, Hornell: Operationism analyzed operationally. *Philos. Sc., 7:*288-313, 1940.
48. ———: Toward an operational definition of the term "operation." *Am. Soc. Rev., 18:*612-7, 1953.
49. Hempel, C. G.: *Fundamentals of Concept Formation in Empirical Science.* Chicago, Univ. Chicago, 1952. (Int. Ency. of Unified Science, Vol. II, no. 7.)
50. Hobson, E. W.: *The Domain of Natural Science.* New York, Macmillan, 1923.
51. Lenzen, Victor: Operational theory in elementary physics. *Am. Phy. Teacher, 7:*367-72, 1939.
52. ———: *Physical Theory.* New York, Wiley, 1931.
53. Lindsay, R. B.: Critique of operationism in physics. *Philos. Sc., 4:*456-70, 1937.
54. ———, and Margenau, H.: *Foundations of Physics.* New York, Wiley, 1936.
55. Locke, John: *An Essay Concerning Human Understanding.* Oxford, Clarendon, 1894.
56. Lundberg, G. A., *et al.: Trends in American Sociology.* New York, Harper, 1929.
57. ———: Concept of law in the social sciences. *Philos. Sc., 5:*189-203, 1938.
58. ———: Operational definitions in the social sciences. *Am. J. Sociol., 47:*727-43, 1942.
59. ———: *Foundations of Sociology.* New York, Macmillan, 1939. Courtesy of the publisher.
60. ———: *Social Research.* 2nd Ed. London, Longmans Green, 1942.
61. Margenau, Henry: Methodology of modern physics. *Philos. Sc., 2:*48-72, 164-87, 1935.
62. ———: *The Nature of Physical Reality.* New York, McGraw-Hill, 1950.

63. Marx, Melvin: *Psychological Theory*. New York, Macmillan, 1951. Courtesy of the authors, the *Psy. Rev.* and the Am. Psy. Assn.
64. Morris, Charles: *Signs, Language, and Behavior*. New York, Prentice-Hall, 1946.
65. Nagel, Ernest: Operational analysis as instrument for the critique of linguistic signs. *J. Philos.*, 39:177-89, 1942.
66. Newton, Isaac: *Mathematical Principles of Natural Philosophy*. Berkeley, Univ. Calif., 1934.
67. ———: *Optiks*. 4th ed. London, G. Bell, 1931. Courtesy of the publisher.
68. Northrup, F. S. C.: *The Logic of the Sciences and the Humanities*. New York, Macmillan, 1947.
69. Peirce, C. S.: *Collected Papers*. Edited by Charles Hartshorne and Paul Weiss. Cambridge, Harvard University, 1932.
70. Porterfield, Austin L.: *Creative Factors in Scientific Research*. Durham, Duke Univ., 1941.
71. Pratt, C. C.: *Logic of Modern Psychology*. New York, Macmillan, 1939.
71.1 *Present State of Operationalism*. Sc. Monthly, 79:209-31, 1954. Contains contributions by H. Margenau, Gustav Bergmann, Carl G. Hempel, R. B. Lindsay, P. W. Bridgman, Raymond J. Seeger and Adolf Grünbaum. This valuable symposium appeared too late to be included in the source material. Bridgman's contribution to it (30) was available in manuscript.
72. Rapoport, Anatol: *Operational Philosophy*. New York, Harper, 1953.
73. Reichenbach, Hans: *Experience and Prediction*. Chicago, Univ. Chicago, 1938.
74. Roback, A. A.: *History of American Psychology*. New York, Library Publishers, 1952.
75. Runes, D. D., (ed.): *Dictionary of Philosophy*. Philos. Lib., New York, 1942, article on Operationism.
76. Russell, Bertrand: *Mysticism and Logic*. London, Longmans Green, 1921.
77. ———: *Introduction to Mathematical Philosophy*. New York, Macmillan, 1920.
78. ———: *Principles of Mathematics*. 2 ed. New York, Norton, 1938.
79. ———: *Inquiry Into Meaning and Truth*. New York, Norton, 1940.
80. Skinner, B. F.: *Behavior of the Organism*. New York, Appleton-Century, 1938.
81. ———: *Science and Human Behavior*. New York, Macmillan, 1953.

82. Stevens, S. S.: The operational basis of psychology. *Am. J. Psy.*, 47:323-30, 1935.

83. ————: The operational definition of psychological concepts. *Psy. Rev.*, 42:517-27, 1935. Courtesy of the editors and of the Am. Psy. Assn.

84. ————: Psychology: the propaedeutic science. *Philos. Sc.*, 3:90-103, 1936.

85. ————, (ed.): *Handbook of Experimental Psychology.* New York, Wiley, 1951.

86. *Symposium on Operationism. Psy. Rev.*, 52:241-94, 1945. Contains contributions by E. G. Boring, P. W. Bridgman, Herbert Feigl, Harold E. Israel, Carroll C. Pratt, and B. F. Skinner. Courtesy of the editors and of the Am. Psy. Assn.

87. Tolman, E. C.: *Purposive Behavior in Animals and Men.* New York, Century, 1932.

88. ————: Discussion from symposium. *J. Personality*, 18:48-50, 1949.

89. Vaihinger, Hans: *Philosophy of "As If."* New York, Harcourt, Brace, 1925.

90. The pervasiveness of the problem of the openness of symbols can only be indicated here. In its broadest formulation it includes such problems as the irreducibility of universals to particulars, of objects to "appearances," of mind to behavior, of "dispositional predicates" to the ordinary predicates, of laws to instances (the inductive problem), of *Gestalten* to their elements, and of the past or future to the present. In this connection the introduction by the logical positivists of the notion of "incomplete reducibility" (Carnap: Testability and meaning, *Philos. Sc.*, 3:420-71, 1936; 4:2-40, 1937) seems to imply a recognition of openness of meaning. The advocates of "emergence" ("emergent evolution," "emergent naturalism") appear to be attempting to call our attention to another kind of openness. References to the problem, other than those already mentioned, are the following: Egon Brunswik (*31*:44; this work also contains a bibliography); J. B. Conant (*Proc. Am. Acad. Arts & Sc. 80*:1, p. 12); G. S. Lundberg: *60*:92-3; F. S. C. Northrup: 68:130; and Frederich Waismann (Verifiability. *Essays on Logic and Language*, ed. by Antony Flew, Philos. Lib., New York, 1951, p. 119).

INDEX

This Book

OPERATIONISM

By A. Cornelius Benjamin

was set, printed and bound by the E. W. Stephens Publishing Company of Columbia, Missouri. The page trim size is 5½ x 8½ inches. The type page is 23 x 39 picas. The type face is Caledonia set 11 point on 13 point. The text paper is 60-pound Mountie Eggshell. The cover is Roxite LS Vellum, 5175, 11-M, Two-tone Black.

With THOMAS BOOKS *careful attention is given to all details of manufacturing and design. It is the Publisher's desire to present books that are satisfactory as to their physical qualities and artistic possibilities and appropriate for their particular use.* THOMAS BOOKS *will be true to those laws of quality that assure a good name and good will.*